AF759705

Principles of Alluvial Fan Morphology

Dan Bowman

Principles of Alluvial Fan Morphology

Springer

Dan Bowman
Department of Geography and Environmental Development
Ben-Gurion University of the Negev
Beer-Sheva, Israel

Front cover: Aerial view of the distal front of the Ze'elim alluvial fan, prograding toward the receding Dead Sea, the Rift Valley, Israel. The drainage pattern changes from diverging and aggrading at the alluvial fan front (lower part of photo) to a converging and entrenching dendritic system on the exposed coastal area (the upper part of photo). (Map data Google Earth, DigitalGlobe, CC BY-4.0)

ISBN 978-94-024-1556-8 ISBN 978-94-024-1558-2 (eBook)
https://doi.org/10.1007/978-94-024-1558-2

Library of Congress Control Number: 2018962002

© Springer Nature B.V. 2019
This work is subject to copyright. All rights are reserved by the Publisher, whether the whole or part of the material is concerned, specifically the rights of translation, reprinting, reuse of illustrations, recitation, broadcasting, reproduction on microfilms or in any other physical way, and transmission or information storage and retrieval, electronic adaptation, computer software, or by similar or dissimilar methodology now known or hereafter developed.
The use of general descriptive names, registered names, trademarks, service marks, etc. in this publication does not imply, even in the absence of a specific statement, that such names are exempt from the relevant protective laws and regulations and therefore free for general use.
The publisher, the authors and the editors are safe to assume that the advice and information in this book are believed to be true and accurate at the date of publication. Neither the publisher nor the authors or the editors give a warranty, express or implied, with respect to the material contained herein or for any errors or omissions that may have been made. The publisher remains neutral with regard to jurisdictional claims in published maps and institutional affiliations.

Cover illustration: Image: Google, DigitalGlobe © 2017

This Springer imprint is published by the registered company Springer Nature B.V.
The registered company address is: Van Godewijckstraat 30, 3311 GX Dordrecht, The Netherlands

This volume is devoted to my mother Trude Schneider Bowman and to my father Dr. Robert Schneider from Eger (Cheb) Czechoslovakia, who was killed by the British Navy when entering Palestine on board of the refugee vessel "Tiger Hill" and to my stepfather Dr. Ernst Bowman, to whom I owe my education. We were all escaping the Nazi regime.

Preface

Alluvial fans have been studied and restudied by many researches worldwide, amounting to an inventory of hundreds of publications and more, many of which have been studied while preparing this volume. The challenge was to bring together the knowledge dispersed widely in the literature, including regional case studies. The manifold aspects addressed in the present volume – morphology, processes, systems, ratios, rates, temporal aspects, controlling factors, and human impacts – should provide a useful basis for understanding alluvial fans. A selection of papers has been attached to each chapter for additional, more focused reading. The complex interdependence of the fan morphology and the mechanism of sediment transport and distribution along confined channelized flows or unconfined surfaces needs, for example, additional reading.

This volume is intended to serve students in earth sciences, engineers, and planners. Of significant help was my experience in lecturing the Alluvial Fan course at Ben-Gurion University of the Negev, Israel, including field trips to the Dead Sea rift with its multiple examples of alluvial fans in different stages of evolution.

Efforts have been made in the past to summarize various aspects of alluvial fans. This volume is an attempt to submit the principles of the great bulk of material in a textbook mode. The following are the main former reviews, summary papers, conference proceedings, and special volumes devoted to alluvial fans:

Blair TC, McPherson JG (1994) Alluvial fans and their natural distinction from rivers based on morphology, hydraulic processes, sedimentary processes, and facies assemblages. J Sediment Res 64A:450–489

Blair TC, McPherson JG (2009) Processes and forms of alluvial fans. In: Parsons AJ, Abrahams AD (eds) Geomorphology of desert environments, 2nd edn. Springer, New York, pp 413–467

Bull WB (1977) The alluvial fan environment. Prog Phys Geogr 1(2):227–270

Clark LE (2015) Experimental alluvial fans: advances in understanding of fan dynamics and processes. Geomorphology 244:135–145

Ethridge, FG (1985) Modern alluvial fans and deltas in recognition of fluvial depositional systems and their resource potential. In: Flores RM, Ethridg FG, Miall AD, Galloway WE, Fouch TD (eds), SEPM short course. Tulsa, SEPM, 19, pp. 101–143

French RH (1987) Hydraulic processes on alluvial fans. Developments in Water Science 31, Elsevier, Amsterdam

Giles P, Nichols G, Wilford D (eds) (2010) Alluvial fans: from reconstructing past environments to identifying contemporary hazards. Geomorphology 115(3–4)

Harvey AM (2011) Drylands alluvial fans. In: Thomas DSG (ed) Arid Zone Geomorphology: process, form and change in drylands, 3rd edn. Wiley, Chichester, pp. 333–371

Harvey AM, Mather AE, Stokes M (2005) Alluvial fans: Morphology, Sedimentology, Dynamics. Geological Society, London, Special publications vol. 251, London

Nilsen TH (1985) Modern and ancient alluvial fan deposits. Benchmark papers in geology series, Van Nostrand Reinhold, New York, 372 p

Nilsen TH, Moore TE (1984) Bibliography of alluvial-fan deposits. Geo Books, Norwich

Rachocki AH, Church M (1990) Alluvial fans – a field approach. Wiley, Chichester, 391 p

Stock JD (2013) Waters divided: a history of alluvial fan research and a view of its future. In: Shroder JF, Wohl E (eds) Treatise on geomorphology, vol 9, fluvial geomorphology. Academic Press, Elsevier, San Diego

Beer-Sheva, Israel Dan Bowman
E-mail: dbowman@bgu.ac.il

Acknowledgement

Thanks go to Ithamar Perath for his constructive reviews on the earlier versions of the manuscript and to two anonymous reviewers. Mrs. Devorah Kremer is acknowledged for producing the illustrations. Funds for preparing the manuscript were granted by the Ben-Gurion University, Beer-Sheva, Israel.

Thanks to Ruth, my friend for life, who brought the desert morphology into our home by her aquarelle paintings which I, Michal, Ehud, and Tomer endlessly enjoyed.

I am grateful for the permissions granted to all the republished figures.

Dan Bowman

Contents

Definitions and Setting

1

Abstract

The alluvial fan is a sediment storage in form of a prism by the mountain front, above a local base-level. Fans show in plan-view a triangular cone-shaped form, bridging in between highly erosive upland catchments and down-fan valleys. To form fans requires (a) an elevational difference and (b) a sharp change in the alluvial geometry when exiting the mountain front. Fan deltas are subaerial alluvial fans that prograde into a standing water body often as Gilbert-type fan deltas, recognizable by a tripartite internal geometry. Distinction should be made between mountain-front fans and tributary-junction fans.

Keywords

Playa · Fluvial geometry · Base-level · Triangular facet · Bajada · Apex · Fan embayment · Mid-fan · Toe · Pediment · Terminal fan · Accommodation space · Fan delta · Gilbert-type fan delta · Mountain-front fan · Tributary-junction fan

Drew created the term *alluvial fan* in 1873. The alluvial fan morphology bridges between the highly erosive upland catchments, where channels are confined and the sediment is produced and the down-fan basins or valley floors where channels can spread and the material is finally deposited. The classic alluvial fan type is a three-dimensional distributive stream deposit at the mountain feet, below a front escarpment (Fig. 1.1), piled up and stored on the way down to an ultimate sink. The initial condition for depositing alluvial fans requires *elevational differences* as between mountains and adjacent intermontane basins which operate as an *accommodation space* for sediment accumulation. A fault system may control the accommodation space and cause tilting and differential subsidence of the basin. Rise of a marine or a lacustrine base-level will likewise create an accommodation for alluvial sedimentation. Fans typically veneer mountain fronts of internally drained closed basins which are tectonic troughs, often with a salt terminal lake or with an extreme flat valley floor called *playa*.

© Springer Nature B.V. 2019

D. Bowman, *Principles of Alluvial Fan Morphology*,
https://doi.org/10.1007/978-94-024-1558-2_1

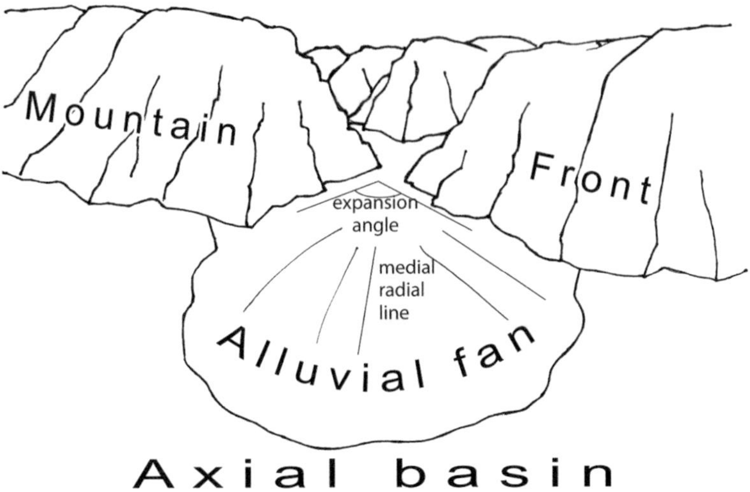

Fig. 1.1 An alluvial fan located at the feet of a mountain front

Each fan has its upstream *sourcing drainage basin* from which water and sediments are funneled through a confined mountain stream, whether perennial, intermittent, or ephemeral. From the point where the channel issues toward the valley, flows expand and become free to swing laterally, to braid and to form a distributive triangular network. Such branching systems are typically associated with environments of net deposition. The classical alluvial fan resembles a cone with an expansion angle, radiating downslope from a sharp termination of the mountainous relief, making a convex pattern of downslope bowing contours. Debris flows, landslides, and talus cones often demonstrate similar conical fan-shaped landforms. A *medial radial line* often splits fans into two halves. Because of confinement problems along their boundaries, not all alluvial fans are semi-conical shaped.

Alluvial fans are the outcome of the erosional response to a rising relief. The fans compose a sediment storage in the form of a prism by the mountain front and serve as a periodic buffer for sediments and water in their move from mountainous erosive catchments. The alluvial fan environment is often hundreds of meters above the *local terminal base-level*. The bedrock slopes along the range front display *triangular facets* between the fan outlets, i.e., triangular mountain-front bedrock slopes between the V-shaped valleys (Figs. 1.2 and 1.3). The magnitude of the facets depends mainly on the width of the drainage basins. Between long and narrow drainage basins, the facets are small and closely spaced.

The term *terminal fan* (Tunbridge 1984) has been suggested for fluvial distributary ephemeral systems in arid and semiarid environments, where high evaporation and percolation into the channel bed make a significant downstream reduction in discharge. The flow splits into smaller distributaries and causes runoff to vanish gradually (Kelly and Olsen 1993; Nichols and Fisher 2007). The morphological,

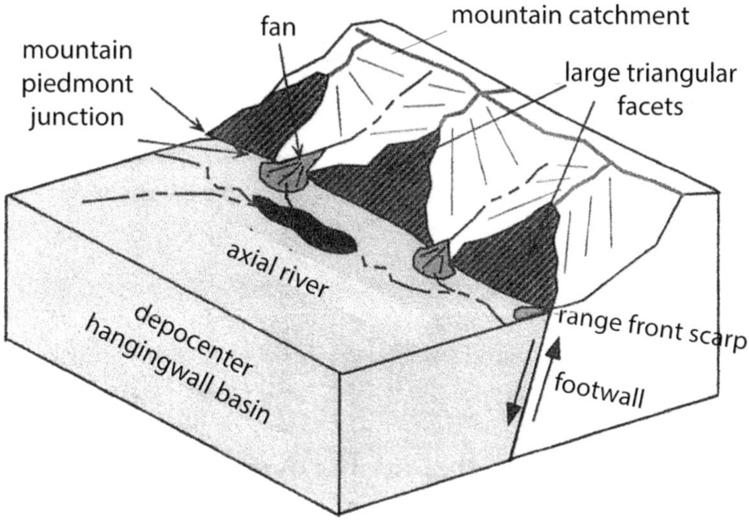

Fig. 1.2 Typical alluvial fans location along a normal fault at a range front. Sourcing drainage basins are entrenched in the mountainous footwall. Triangular facets indicate the mountain front. The fans prograde towards the axial river in the hanging wall basin. (Modified after Burbank and Anderson 2011)

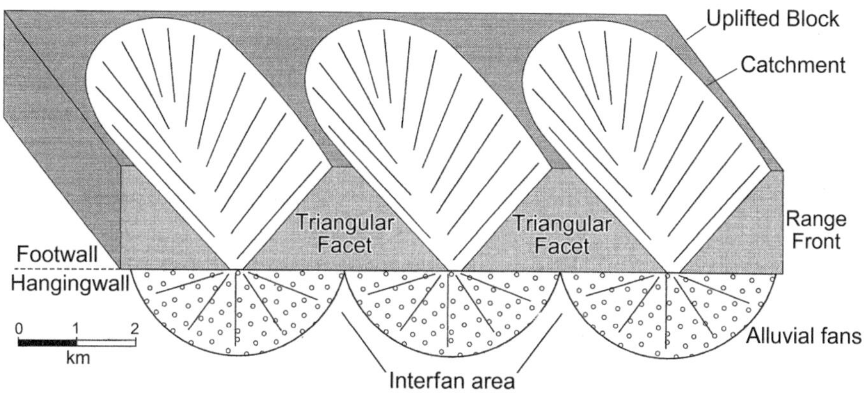

Fig. 1.3 Typical alluvial fan location along a range front. The triangular facets may retain partly the initial fault plane. They owe their size to the spacing of the channels. (Modified after Blair 1999)

sedimentological, stratigraphical, and dynamic characteristics which have been related to terminal fans (Friend 1978; Olsen 1987; Abdullatif 1989; Kelly and Olsen 1993; Friend et al. 1999; Arzani 2007; Stephen and Mountney 2009) are those typical to alluvial fans. The terminal fan concept does not contribute to an improved understanding of the alluvial fan environment. It seems unjustified to use the *terminal fan* concept as a distinctive fan model, a conclusion also shared by North and Warwick (2007).

Alluvial fan evolution is often triggered by thrusting (Fig. 1.4). Deformation pulses may uplift and rejuvenate the source areas. Long periods of mountain uplift and sinking of the nearby basins are important for maintaining the elevational differences and allowing continuous fan sourcing. Fan progradation is controlled by the ratio of subsidence of the accommodation space to the sediment supply from the hinterland. If the rate of sediment delivery and the flux from the source basin exceeds the subsidence, fans will overflow and prograde (Clevis et al. 2004). Fans may prograde tens of kilometers away from the active thrust front. Increase in accommodation space by subsidence will confine the accumulation of fans closer to the mountain front.

Alluvial fans can be distally confined by an axial river and laterally by neighboring fans or stay unconfined (Fig. 1.5). Alluvial fans may thus appear individually uncoalesced as a well-defined single unit with interfan areas drained by channels sourced from the mountain front. Fans may alternatively show lateral

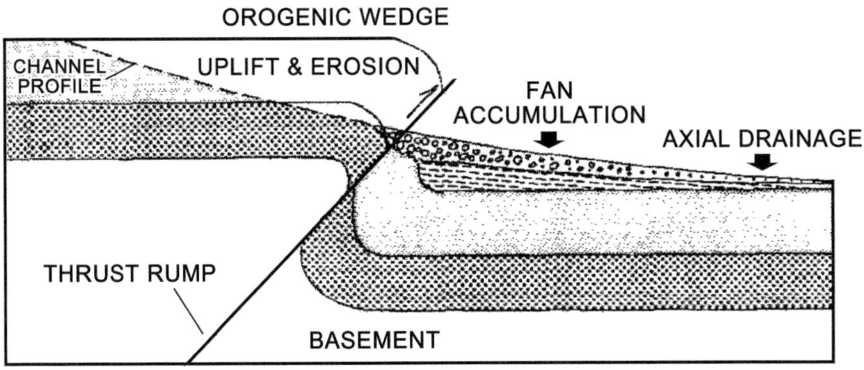

Fig. 1.4 The location of alluvial fans in front of a foreland uplift following unroofing. The uplifted orogenic wedge above the thrust ramp includes the sourcing drainage basins. (Modified from Decelles et al. 1991)

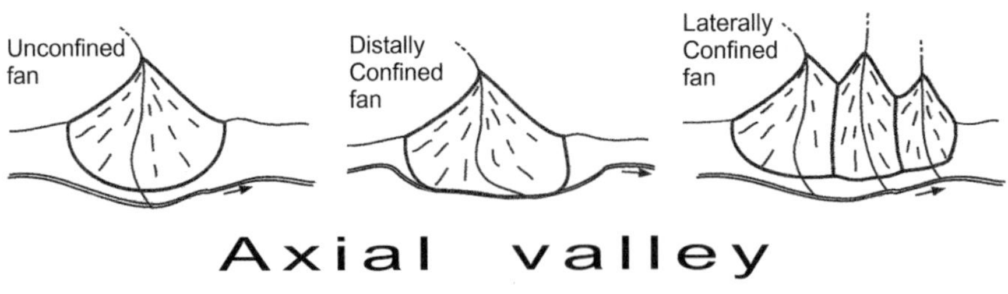

Fig. 1.5 Categorization of alluvial fan confinement. (After Crosta and Frattini 2004)

coalescing with neighboring fans or with the axial basin deposits. Closely spaced alluvial fans may interfinger but not merge, each remaining an individual structure. The narrower the spacing between the catchment outlets, the more the fans become coalesced. Fully merged alluvial fans are called *bajada* (Eckis 1928) and lack a fan-shaped form. It may attain lengths of over 100 km and a thickness of a few tens to a few hundreds of meters, composing a continuous depositional apron. The convexity of its individual fan cones is lost, and contour lines become parallel and linear (Blair and McPherson 2009). Topographically, alluvial fans display typical half-circular contours and show typical convex cross profiles parallel to the mountains (Figs. 1.6 and 1.7) with the central part of the fans almost invariably higher because of more deposition (Hooke and Rohrer 1979). The typical

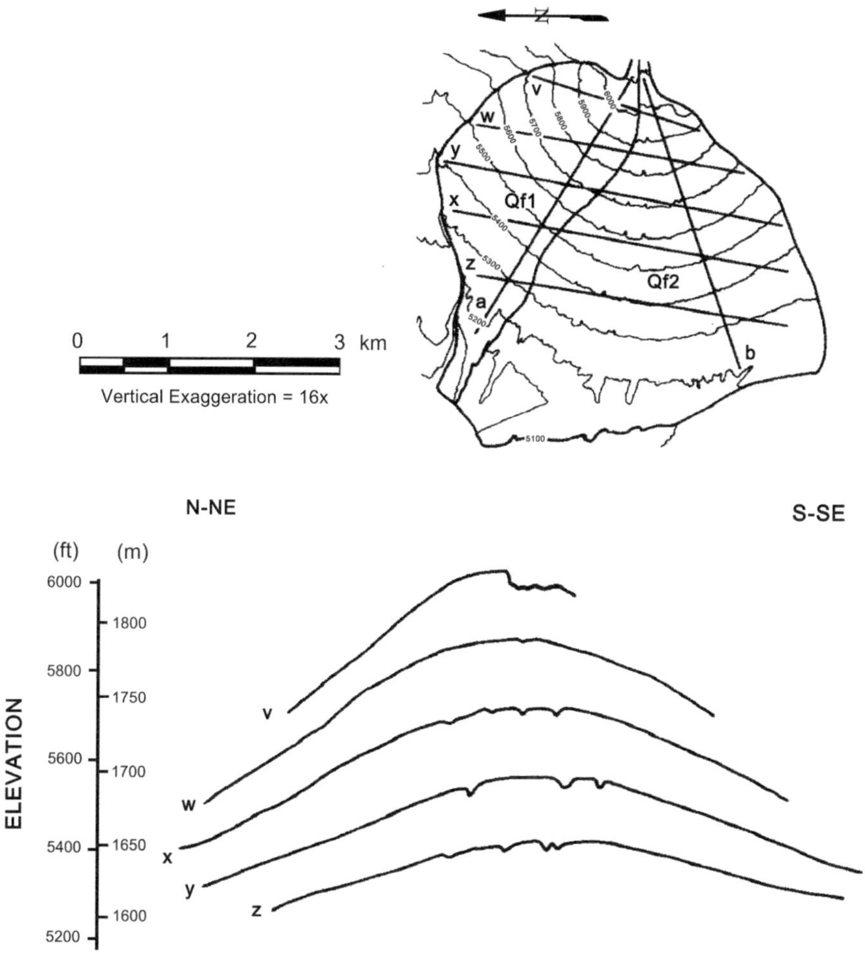

Fig. 1.6 Characteristic convex cross-fan profiles of Cedar Creek alluvial fan. (Following Ritter et al. 1993)

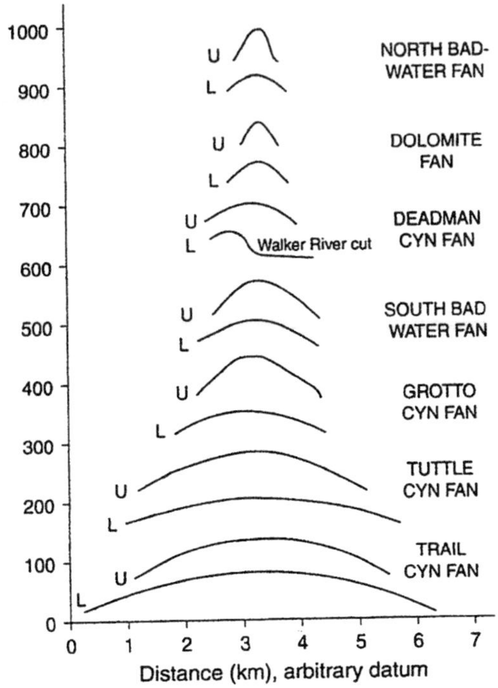

Fig. 1.7 Cross profiles of fans from California, including the Death Valley and Owens Valley. Note the convexity of the fans at the upper apex fan area (U) and the flatter form at the lower (L) area of the fans. (From Blair and McPherson 2009). The steeper lateral slopes invite avulsion. Channels will finally change their path and follow the steeper lateral descent

down-fan decrease in the cross-fan convexity is well illustrated by the progressive straightening of the contour lines toward the toe.

The *fan head* refers to the area around the *fan apex*, the highest point on the alluvial fan from where the surface spreads downslope. This area is generally of a higher energy relative to the *mid-fan* and the *distal areas*. When material backfills the channel *upstream* from the mountain front, a *fan bay* is formed (Harvey 2011, Figs. 1.8 and 1.9). The embayments filled up with sediments make a characteristic sharp, teeth-like, irregular mountain front. The *mid-fan* is a more stable zone without extreme trends. In many cases the mid-fan composes most of the fan area and could be further divided into the upper and lower mid-fan. The span from the lower mid-fan to the *toe* serves mainly as the zone of deposition. The *base* of the fan refers to the lowest, outermost fan area. The distal area of alluvial fans may grade into the top surface of an axial river terrace or a floodplain. Axially derived fluvial deposits may be separated from the transversely derived alluvial fans, making two different sets of mineralogical compositions.

Pediments encircle the upland relief above the alluvial lowlands (Dohrenwend 1994). Pediments are nearly planar, gently sloping (1–6°) rocky mountain flanks, generally thinly veneered with weathered colluvium or covered by a deep weathering mantle up to a few meters thick but may also be bare. Morphologically, the pediment makes a flat, featureless, overall undissected, low-relief (<1 m) surface, often discordant to the structure. Along the pediments can often be observed an upper

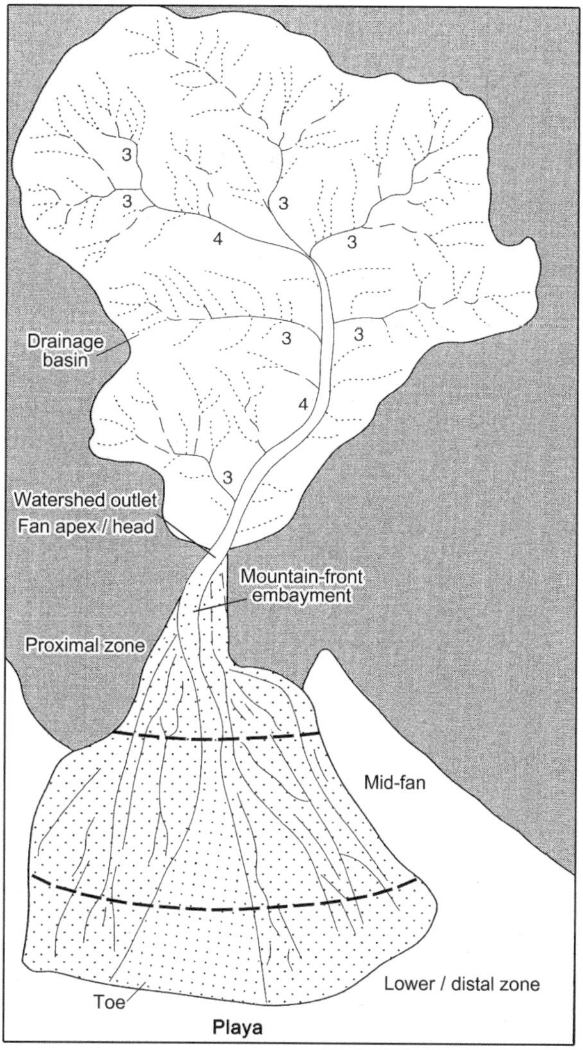

Fig. 1.8 A typical alluvial fan system composed of the fan and its zones and the sourcing catchment within the mountainous bedrock, including the channel orders. (Modified after Blair and McPherson 1994)

erosional zone, a medial transportation zone, and a low depositional area. *Playa* is a periodically flooded area in the axial basin, in front and down of alluvial fans. It is the lowest-lying area and thus a local base-level (Fig. 1.8). The playa may become an ephemeral lake that may completely dry up. The playa stratigraphy will characteristically show layers of anhydrite, halite, and gypsum with fine-grained detritus that represent the suspended fine fraction of the downstream floods, often intercalated with sand and some gravel.

Chronologically, a distinction is made between ancient fans, quaternary fans, and modern alluvial fans. Another differentiation is made between *mountain-front fans*, which make up the main classic/conventional fan category, and the *tributary-junction fan* type formed in a confined valley setting where a tributary valley joins a

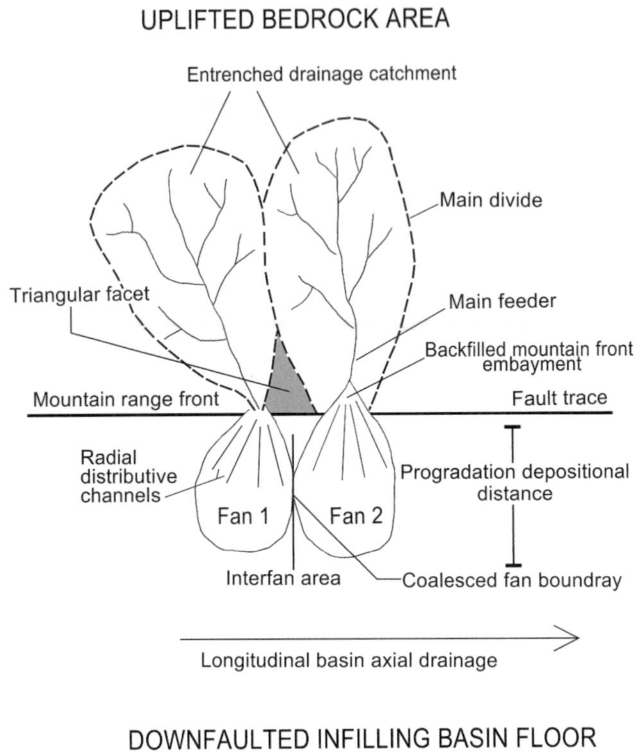

Fig. 1.9 The alluvial fan environment

larger trunk valley. Spatial confinement of tributary-junction fans depends on the position of the main river on the valley floor, i.e., how much space is available for deposition of the tributary-junction fan. The proximity of the trunk river channel to the tributary fan, together with their sediment/water discharge relationships, are key controlling factors. Only if the power of the trunk river is less than needed to remove the material supplied by the tributary, deposition can be expected at the junction (Wang et al. 2008). Small tributaries with limited sediment supply may not generate tributary-junction fans when the power of the main river may be sufficient to remove all the sediment supplied by the tributary. When formed, the tributary fan may be deposited on top of the valley floor or fill incision within the valley fill. Tributary confined fans may temporarily block the trunk valley before being breached (Stokes and Mather 2015). Another fan category develops at non-tectonic sidewalls of deglaciated U-valleys, where a postglacial tributary channel joins the deep, scoured, former glacial valley.

Fan deltas are subaerial alluvial fans that prograde into a ponding water such as a lake or sea (McPherson et al. 1987; Massari and Colella 1988; Nemec and Steel 2009; Dart et al. 1994). The level of the water body is a major control of fan deltas, and their coastal front is typically arcuate with lobes at the river mouths (Schumm et al. 1987). The fan deltas are very different from *deltas* that are typically not located close to exits from a mountain front, are composed of finer and sorted deposits, and have very smooth slopes. Fossil shoreline features on subaerial fans indicate that they were originally fan deltas. The fan deltas are usually composed of:

(a) *Alluvial fan* deposits which compose the subaerial proximal and mid-fan environments and comprise, in most cases, the largest part of the fan delta.
(b) A *transition zone* where the fluvial environment interfingers with coastal and littoral processes. Recognition of the coastal facies is possible through its textural characteristics, i.e., good sorting, high roundness, seaward imbricated gravel, swash lamination, horizontally bedded berm sands, or landward dipping backshore deposits.
(c) The *subaqueous zone* of the fan delta often covers an area a few times larger than the subaerial fan. Both aggradation and progradation increase the size of fan deltas.

Fan deltas may compose, although not always, classical *Gilbert-type fan deltas*, marked by a *tripartite internal geometry* (Gilbert 1885; Fig. 1.10) which consists of (**a**) *topset beds* of a fluviatile origin, often poorly sorted, crudely horizontally bedded with landward imbricated conglomerates. This subaerial to shallow marine alluvial component overlies the foresets below. (**b**) The prograding fan-delta front is composed of steeply 25–35° dipping *foreset*s or *clinoforms*, exhibiting slide surfaces, avalanche deposits, and retrogressive slumping. This steep fan-delta front slope is a major marker of fan deltas. (**c**) The *bottomsets/toesets* of the *prodelta* are gently dipping (<10°) at the deeper fan-delta front, composed of siltstones and sandstones. The bottomsets settle in the quiet deeper water and are composed of

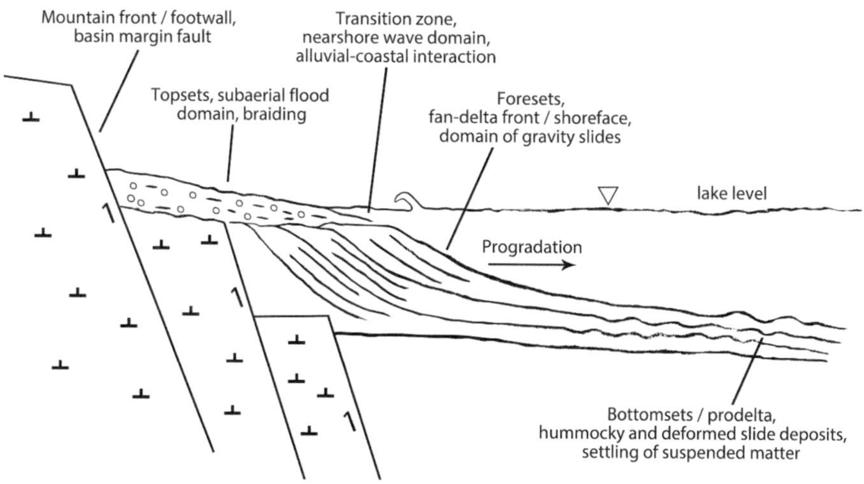

Fig. 1.10 Schematic section through a Gilbert-type fan delta showing its spatial facies distribution

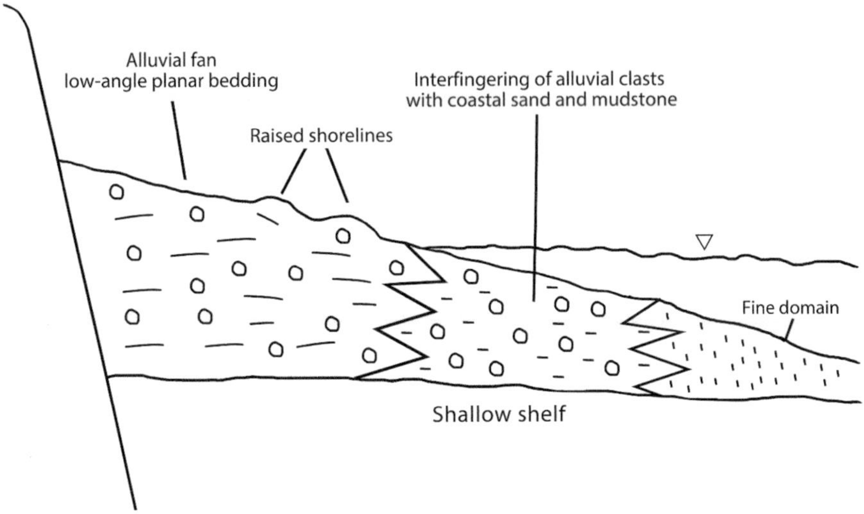

Fig. 1.11 Schematic section through a shelf-type fan delta showing its spatial facies distribution

suspended load, winnowed by wave action from the nearshore surf zone, or are partly removed by sliding of foresets (Potsma and Roep 1985).

Gilbert-type fan deltas have a radius of several km, may be up to hundreds of m thick, and usually develop along steep fault scarps at basin margins. Rapid basin subsidence or hinterland uplift enhances the *vertical stacking* of successive superimposed Gilbert-type fan deltas and triggers less basinward progradation. On a gentle sloping physiography bordering a shallow water basin, we expect a *shelf model* fan delta to develop which is a *low-gradient fan-delta* type (Fig. 1.11). Different paleo-fan-delta types facilitate reconstruction of ancient basins. Hydrocarbon-bearing fan-delta deposits have been found along plate margins. The current interest in fan deltas stems largely from their potential as hydrocarbon reservoirs.

References

Abdullatif OM (1989) Channel-fill and sheet-flood facies sequences in the ephemeral terminal river gash, Kassala, Sudan. Sediment Geol 63:171–184

Arzani N (2007) Playa-lake level fluctuation and recent sediments of a terminal fan-playa fringe, Abarkoh Basin, Central Iran. J Sci Islam Repub Iran 18(1):19–33

Blair TC (1999) Alluvial fan and catchment initiation by rock avalanching, Owens Valley, California. Geomorphology 28:201–221

Blair TC, McPherson JG (1994) Alluvial fans and their natural distinction from rivers based on morphology, hydraulic processes, sedimentary processes and facies assemblages. J Sediment Res 64:450–489

Blair TC, McPherson JG (2009) Processes and forms of alluvial fans. In: Parsons AJ, Abrahams AD (eds) Geomorphology of desert environments, 2nd edn. Springer, New York, pp 413–467

Burbank DW, Anderson RS (2011) Tectonic geomorphology. Wiley-Blackwell, Oxford

Clevis Q, De Boer PL, Nijman W (2004) Differentiating the effect of episodic tectonism and eustatic sea-level fluctuations in foreland basins filled by alluvial fans and axial deltaic systems: insights from a three- dimensional stratigraphic forward model. Sedimentology 51:809–835

Crosta GB, Frattini P (2004) Controls on modern alluvial fan processes in the Central Alps, northern Italy. Earth Surf Process Landf 29(3):267–293

Dart CJ, Collier REL, Gawthorpe RL, Keller JV, Nichols G (1994) Sequence stratigraphy of (?) Pliocene-Quaternary synrift, Gilbert-type fan deltas, northern Peloponnesos, Greece. Mar Pet Geol 11(5):545–560

Decelles PG, Gray MB, Ridgway KD, Cole RB, Srivastava P, Pequera N, Pivnik DA (1991) Kinematic history of a foreland uplift from Paleocene synorogenic conglomerate, Beartooth Range, Wyoming and Montana. Geol Soc Am Bull 103:1458–1475

Dohrenwend JC (1994) Pediments in arid environments. In: Abrahams AD, Parsons AJ (eds) Geomorphology of desert environments. Chapman & Hall, London, pp 321–353

Drew F (1873) Alluvial and lacustrine deposits and glacial records of the Upper Indus Basin. Geol Soc London Q J 29:441–471

Eckis R (1928) Alluvial fans in the Cucamonga district, southern California. J Geol 36:224–247

Friend PF (1978) Distinctive features of some ancient river systems. In: Miall AD (ed) Fluvial Sedimentology. Canadian Society of Petroleum Geologists, Calgary, Memoir 5

Friend PF, Jones NE, Vincent SJ (1999) Drainage evolution in active mountain belts: extrapolation backwards from present-day Himalayan river patterns. In: Smith ND, Rogers J (eds) Fluvial sedimentology. Sp Publ Int Ass Sediment 28: 305–313

Gilbert GK (1885) The topographic features of lake shores. Ann Rept US Geol Surv 5:69–123

Harvey A (2011) Dryland alluvial fans. In: Thomas DSG (ed) Arid Zone Geomorphology: process, form and change in drylands, 3rd edn. Wiley, Chichester, pp. 333–371

Hooke RL, Rohrer WL (1979) Geometry of alluvial fans: effect of discharge and sediment size. Earth Surf Process Landf 4:147–166

Kelly SB, Olsen H (1993) Terminal fans – a review with reference to Devonian examples. Sediment Geol 85:339–374

Massari F, Colella A (1988) Evolution and types of fan-delta systems in some major tectonic settings. In: Nemec W, Steel RJ (eds) Fan deltas: sedimentology and tectonic settings. Blackie, Glasgow, pp 103–122

McPherson JG, Shanmugam G, Moiola RJ (1987) Fan-deltas and braid deltas varieties of coarse-grained deltas. Geol Soc Am Bull 99:331–340

Nemec W, Steel RJ (eds) (2009) Fan deltas: sedimentology and tectonic settings. Blackie, Glasgow

Nichols GJ, Fisher JA (2007) Processes, facies and architecture of fluvial distributary system deposits. Sediment Geol 195:75–90

North CP, Warwick GL (2007) Fluvial fans: myths, misconceptions and the end of the terminal-fan model. J Sediment Res 77:693–701

Olsen H (1987) Ancient ephemeral stream deposits: a local terminal fan model from the Bunter sandstone formation (L. Triassic) in the Tonder–3, –4 and –5 wells, Denmark. In: Frostick L, Reid I (eds) Desert sediments: ancient and modern, vol 35. Geological Society Special Publication, pp 69–86

Potsma G, Roep TB (1985) Resedimented conglomerates in the bottomset of a Gilbert-type gravel delta. J Sediment Petrol 55:874–885

Ritter JB, Miller JR, Enzel Y, Howes SD, Nadon G, Grubb MD, Summa CL (1993) Quaternary evolution of Cedar Creek alluvial fan, Montana. Geomorphology 8(4):287–304

Schumm SA, Mosley MP, Weaver WE (1987) Experimental fluvial morphology. Wiley, New York., Chap. 10, pp 351–365

Stephen AC, Mountney NP (2009) Spatial and temporal evolution of a terminal fluvial fan system: the Permian Organ Rock Formation, South-East Utah, USA. Sedimentology 56:1774–1800

Stokes M, Mather AE (2015) Controls on modern tributary-junction alluvial fan occurrence and morphology: high Atlas Mountains, Morocco. Geomorphology 248:344–362

Tunbridge IP (1984) Facies model for a sandy ephemeral stream and clay playa complex; the Middle Devon Trentishoe formation of North Devon, U. K. Sedimentology 31:697–725

Wang H, Harvey AM, Xie S, Kuang M, Chen Z (2008) Tributary-junction fans of China's Yangtze Three-Gorges valley: morphological implications. Geomorphology 100:131–139

Magnitude and Sourcing

2

Abstract

Fans are part of an alluvial system that includes an upstream basin from where rock waste is stream-transported to the fan. The sediment load may be entirely or partly deposited on the fan but also totally bypass the fan and leak downcurrent to the axial basin system. A variety of factors and processes control the sediment supply from the catchments to the fans, including the catchment slope, remobilization of sediments stored in former evolutionary stages, and aging. Alluvial fans demonstrate a diversity of sizes. Most fans typically range 0.5–15 km in length. Frequent thickness of alluvial fans ranges from a few tenth up to a few hundred meters. Gravitationally slope-controlled fan-shaped colluvial cones and sheets in mini-sourcing drainage system lack the basic fan characteristics and should not be defined as fans. Megafans which are the largest fans on earth and emanate from outlet points at the front of mountainous fold-thrusts may reach above 100 km in length and cover areas above 10,000 km^2.

Keywords

Fan system · Footwall catchment · Trap efficiency · Sediment-limited · Transport-limited · Paraglacial · Radial fan length · Lobe · Losimean · Kosi megafan · Okavango megafan

2.1 Fan Sourcing

Fans are part of an alluvial system which includes, besides the fan itself, an upstream sourcing basin, linked to the fan by a channel (Fig. 1.8). There are also, although not common, composite fans with multiple source areas. Alluvial fans are typically deposited at mountain fronts along the feet of major fault scarps. The source areas of alluvial fans are typically located on a *footwall catchment* where from material is transported across the mountain front toward a depocenter located on the hanging

© Springer Nature B.V. 2019
D. Bowman, *Principles of Alluvial Fan Morphology*,
https://doi.org/10.1007/978-94-024-1558-2_2

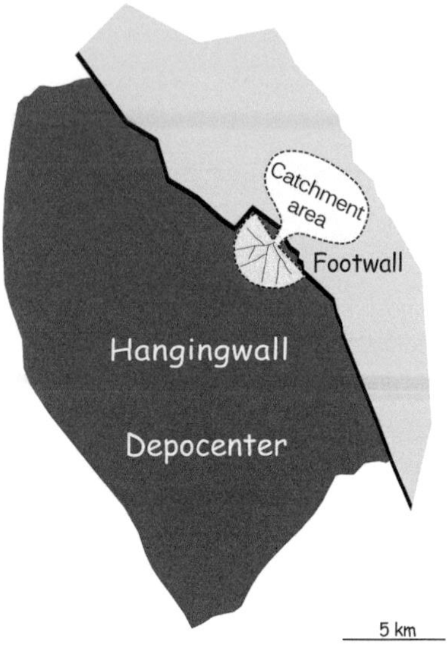

Fig. 2.1 Schematic map showing an alluvial fan system located by a border fault: the catchment is typically located in the mountainous uplifted footwall area (grey) and the fan is found on the down-faulted hanging wall. (Modified after Muravchik et al. 2014)

wall (Figs. 1.2 and 2.1).The sourcing basin may also be located entirely within the hanging wall.

Fans are sourced by debris from the hillslopes of the drainage basin, i.e., by products of weathering that are stream-transported to the fan. The physical and chemical properties of the bedrock in the sourcing basin control the magnitude and the size of the debris to be delivered. In addition, colluvial material may be deposited directly on fan's margins by rockslides along the fault-bounded mountain front. Unique rock avalanche blocks may constitute boulders on the fan surface. Such blocks will be incorporated as outsized angular clasts in the stratified sequence of the fan.

The *trap efficiency* of a fan (Whipple and Trayler 1996) is expressed by the total volume of sediment yield that is deposited on the fan versus the material that bypassed the alluvial fan to be transported down, out of the fan system. Fan aggradation is rapid when the entire sediment load is trapped and deposited on the fan without leakage downcurrent to the axial drainage system. Sooner or later, however, bypassing will develop, and part of the sediments will be transported through the fan further down. The volumetric rate of fan deposition (V_f) is

$$V_f = E_{to}\left(1-\lambda_r / 1-\lambda_s\right) U_r A1_d^{-} r \quad (2.1)$$

where E_{to} is the trap efficiency of one square km of catchment area; r is a constant which, in principle, may vary with process; λ_s is the porosity of the fan sediments; λ_r denotes the rock porosity of the catchment area; U_r is the uplift rate; and A_d is the size of the catchment area. In steady-state conditions lithological and climatic influences are felt through the trap efficiency factor.

Deposition of fans is concurrent with downcutting in the sourcing drainage basin and in reducing its relief. The following processes increase sediment supply from the sourcing drainage basin and promote aggradation of alluvial fans: plenty of debris on unstable hillslopes of soft rocks; increase in rainfall intensity; and removal of vegetation by man or naturally by aridity, landslides, and rockfalls. Slopes supply huge amounts of debris particularly after fires. Minimal supply of slope debris will occur following an increase in vegetation cover and with an increased and prolonged rainfall regime of low-intensity events. Flood events that flush stored sediments and clean up the drainage basin may postpone debris and sediment supply for the period needed for a new significant debris accumulation.

Alluvial fans may be fed by different-sized catchments and by a variety of bedrock lithologies. With time, as the sourcing catchment ages, more bedrock lithologies will become exposed and will be reflected in the spectrum of mineral debris in the stratigraphic sequence of the fan. Distinctive trends in the relative concentration of minerals in fans reflect changes of exposure in their source area throughout the erosional history (DeCelles et al. 1991). Matching conglomerates and source-terrane compositions may prove useful in clarifying the unroofing history of the source terrane.

During some stages in its history, when the entire sediment load is deposited on the fan without outflow to an axial drainage system beyond, the catchment area plus the fan behave as a *semi-closed system* without leakage. The sourcing province may be *sediment-unlimited*, with ample transport-ready volumes of sediment, in contrast to *sediment-limited* basins, where previous events may have already removed most of the available material. *Transport-limited* basins are those where the available sediment is unexceeded by the power of transport, and significant transfer of sediments is prevented.

The main constraints on fan sourcing are the size of the drainage basin, its rate of weathering, the frequency of transport events, and their power. Expansion of catchments with time will cause a reduction in the number of fans in the region. The slopes of the catchments have a significant influence on the sediment supply. Small (<25 km^2) and steep ($>10\%$) catchments tend to supply more sediments and are typically more associated with debris flows (Harvey 1984; Rickenmann and Schiedl 2013). Inputs from larger catchments typically have a higher water/sediment ratio. The larger drainage basins provide a larger sediment storage capacity, mainly along the highest-order valley, and temporally block parts of the sediments from being transferred. Narrow V-shaped mountain valleys leave little space for sediment storage.

Aging of catchments is accompanied by a progressive reduction in their surface relief, often concomitant with increased soil development, increased infiltration, decreased peak discharge, and a diminishing sediment yield. Sediment supply may, however, be intensified by remobilization of unlithified or semi-lithified sediments stored in former evolutionary stages on the slopes and in the valley bottom. Unlithified volcanoclastic sediments may provide such source material, suitable for fan deposition and unrelated to the rate of catchment denudation.

River capture may change the area of a sourcing catchment and its lithologic characteristics (Mather et al. 2000). An extrabasinal stream capture can be recognized through change in the fan sediments due to the influx of new lithologies

(Mikesell et al. 2010). Lack of lithological fit between the source area and its fan, shown by an abrupt lithological change in the fan's stratigraphy, may be caused by a change in the gravel provenance due to movement along a strike-slip fault. The appearance of a different distinct clast lithology in the fan stratigraphy can also be associated with exposure of a particular source rock following an erosional encounter.

The growth or melting of a glacier within a source area may exert a key control on the fan system. Glaciation produces large amounts of glacial drift. Peak sediment sourcing from glaciers occurs during the glacial retreat when till and moraine sediments become exposed. Such *paraglacial period* is characterized by an accelerated release of sediments into the drainage systems as happened following the Little Ice Age. The remobilized and reworked glacier sediments do not reflect the rate of weathering and the denudation of the sourcing area (Dec 1992). Paraglacial sourcing ceases once the sediment supply of the unconsolidated glacial till is exhausted.

2.2 Dimensions

Alluvial fans come in a diversity of sizes, typified by the *axial radial fan length* distance, i.e., by the longest distance that the fan builds out from the mountain front toward the basin, which is the progradation depositional distance (Fig. 1.9). This distance is determined by the basement slope, by the subsidence rate, by the balance between the rate of sediment and the water supply from the drainage area, and by the losses along the way which control the efficiency of the distributive system. The axial radial fan length of most alluvial fans ranges 0.5–6 km. Longer fans are typically in the order of 5–15 km. Exceptionally long fans may reach a length of a few tens of km but usually not beyond 30–45 km.

Fan deposits are typically wedge-shaped, i.e., thickening toward the sourcing catchment at the mountain front and thinning distally (Fig. 1.4). The topographic irregularities of the underlying terrain at the depositional site affect the thickness of the fans. Frequent thickness of alluvial fans ranges from a few tenth of meters (<100 m) up to a few hundred meters. Fans may, however, also constitute only a thin alluvial veneer of a few meters thickness on a beveled surface. Alluvial fans are thicker where the dips of the basin-bounding faults are steeper. Whereas steep faults are sites of thick and small fans, thin and large fan veneers accumulate by low-angle normal faults.

2.2.1 How Small Can Alluvial Fans Be?

There are half-cone-shaped depositional forms which should not be regarded as alluvial fans. These fan-shaped forms include:

1. Valley-side debris cones, i.e., talus of colluvium by mountain slopes and rockfall scree or grainflows by scarps. Sourcing of such deposits comes from the moun-

tain slopes and from front rocky walls. The cones range steeper than steep alluvial fans, i.e., 12–25° (Blair and McPherson 1994) with some overlap and have the classic semi-conical plan-view fan shape with no sourcing drainage basin. Some sediment movement and sorting have been recorded on such cones, probably gravimetric-triggered.

2. Fan-shaped piles of unsorted debris at exits of steep, short, freshly cut gullies, a few hundreds of meters long. This category does not show the typical relationships of drainage basin area/fan area, drainage basin area/fan gradient, and fan area/fan gradient (Al-farraj and Harvey 2005). In this category are included fan-shaped sand masses <1.0 m thick, up to 250 m long, along flanks of fixed dunes (Talbot and Williams 1978) and along a lee face of modern dunes, 55 m high (Sweeney and Loope 2001).
3. Thin veneers of sands that form depositional fan-shaped sheets on slopes following escape of the runoff from rill micro-channel conditions to a wider slope surface.

Although the above mentioned depositional forms show the diagnostic conical fan attribute, they should not be regarded alluvial fans as they lack the following basic fan characteristics:

(a) They have no real sourcing basin, i.e., their catchments are very small, i.e., < 0.1 km^2 and even <0.01 km^2, and often do not compose even a first-order drainage basin but only a short gullied slope, a few hundred meters long, and a few tenths of meter high (Wells and Harvey 1987).
(b) Many basic fan processes as channel shifting, cutting and filling complex multiphase trenching and aggrading are nonoperative.
(c) The depositional forms are often very small. Even when <0.01 km^2 in area, the accretions were still regarded as alluvial fans (Gómez-Villar et al. 2006). However, they lack evidence for being a multi-event depositional product but look more like a single storm deposit without the typical fan accumulative complexity.
(d) Lack of significant channel transportation characteristics whereas alluvial fans are widely regarded as being deposited, wholly or in part, by the tractive forces of flowing water (Stock 2013). The material consists of rockfall and weathered bedrock, i.e., gravitationally controlled without features of sedimentological maturity.
(e) Lack of the typical gradients of fan slopes. The fanlike depositional forms are characteristically steeper (20–35°).

There is a real doubt if gravitationally and slope-controlled fan-shaped colluvial cones at the base of cliffs and sheets at exit of mini drainage systems on slopes should truly be defined as alluvial fans or debris fans, based mainly on their distinctive cone shape. It looks that the triangular shape, although a basic two-dimensional characteristic of alluvial fans, has been much overestimated when used as a criteria for defining slope deposits as *small alluvial fans*, *debris fans*, or *colluvial fans*. A sourcing drainage basin with a feeder as the conveyor of water and sediments is a basic criteria for defining alluvial fans. Accumulations that are a few tenths or hundreds of meters long, deposited by rills and gullies, poorly sorted or unsorted, texturally angular, without bedding, and often sedimentary structureless, piled together by gravitational sliding or by slope runoff and lacking the basic facies characteristics of fluvial water flow, *belong to the hillslope environment*. It is suggested that such colluvial deposits and sheet fanlike accumulations should be excluded from the alluvial fan category.

Small tributary-junction fans should as well be related to the *quasi alluvial fan family* for their following characteristics (Stokes and Mather 2015): (I) Quantitative analysis of their morphometric properties through regression reveals a weak positive relationship between catchment area and fan area and a weak negative relationship between catchment area and gradient. However, although these morphological relationships are weak, they are statistically valid. (II) Most tributary fan areas are undersized <0.015 km^2. The fans are not able to build to their size in respect to the catchment characteristics because of the confined space available and /or because of erosion and resetting by the trunk drainage. (III) There is no space nor evidence for avulsion, channel shifting, coalescing, and for complex multiphase trenching and aggradation, including inset sedimentary bodies, which make up the typical fan accumulative complexity. Basic characteristics of alluvial fans do not form nor function in such small-scale deposits.

Thus although showing the typical fan shape, small tributary fans of foreshortened trimmed fan form, resulting from the interplay between the sediment and water discharge regime of the ephemeral tributary and the perennial trunk river, do not show many fundamental characteristics of alluvial fan and should, not only because of the scale, be defined by a different suitable term as *lobes* or *cones* which better reflect their specific nonfan characteristics.

2.3 Megafans

Megafan, the largest fan category on earth, is an important component of aggradational continental fills in active subsidence basins. Megafans emanate from outlet points at the front of large mountainous fold-thrust belts such as the Andes and the Himalayas and spread over a large foreland basin such as the Indus-Ganges-Brahmaputra alluvial plain. Megafans are deposited by rivers that undergo moderate to extreme seasonal fluctuations in discharge with acute peakedness. The global distribution of modern megafans is primarily restricted to 15–35° latitude both north and south, corresponding to climatic belts that fringe the tropical climatic zone (Leier et al. 2005).

The megafans are supplied by very large sourcing drainage catchments, usually larger than 50.000 km^2, supplying an average depositional rate range of 1.4–3 mm/a. Megafans extend from the apex downstream to distances up to 250 km and reach widths of more than 100 km, covering areas of 10.000–18.000 km^2 vs. generally

<100 km^2 of most alluvial fans. Valleys entrenched in megafans may be a few kms wide, and for discriminating megafans, satellite data is required.

Megafan's longitudinal topographic transects show smooth, concave-up profiles (Figs. 2.2 and 2.3d). In a transverse view, the profiles are convex-up. Topographically, megafans are very flat with slope gradients of <0.01°. The sediment deposits of megafans are current-laid and devoid of gravity flows. Megafans typically show channel instability and rapid lateral migration. Trunk channel relocations in the apex region are frequent and autocyclic (Chakraborty et al. 2010).

Very large *lobes*, i.e., flattish, depositional components with a rounded front, make up megafans, which typically attain a multi-lobate geometry (Fig. 2.3c). Channels commonly split downstream into two or more distributaries showing systematic downstream reduction in width and discharge, typical of distributary systems. The additional typical features and environments that can be distinguished across a megafan are: active channels, abandoned and underfit channels, floodplains, small lakes, ponds, changes from braiding to meandering, permanent swamp vegetation, an abandoned meander belt, low sinuosity meandering defined as "losimean" (Stanistreet and McCarthy 1993), and aeolian sand dunes.

The large catchment areas that feed megafans usually develop through lateral merging of mountainous drainage basins in the hinterland. Gupta (1997) and Horton and DeCelles (2001) suggested a scenario of sideways gathering of adjacent transverse rivers, merging into a single trunk which crosses the mountain front and serves as the point source for a megafan during the growth of structural topographic barriers

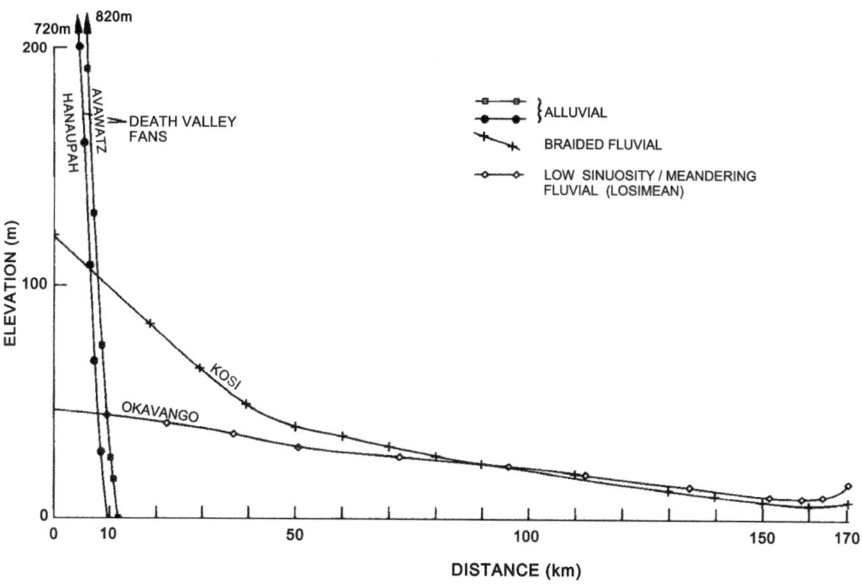

Fig. 2.2 The wide spectrum of down-fan slopes: from small alluvial fans, i.e., <10 km long and hundreds of meter in height, presented by the California Death Valley fans, to megafans more than 150km long and only a few tenth of meter in elevation presented by the Indian Kosi and the Botswana Okavango megafans. (Redrawn from Stanistreet and McCarthy 1993)

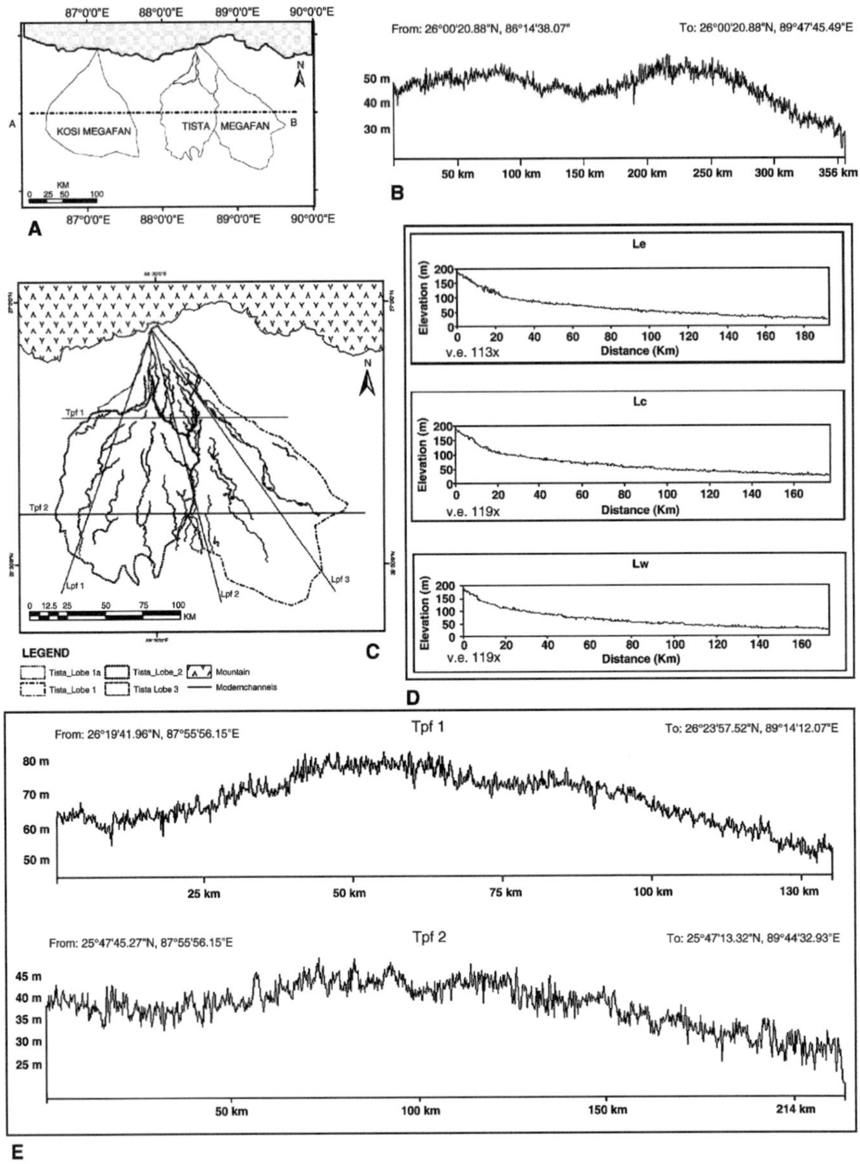

Fig. 2.3 Diagram showing the outline of the Tista and the Kosi megafans at the foothills of Darjeeling Himalayas and the location of the transverse profile (**a**). The cross profile of the Kosi and the Tista megafans is shown on (**b**). The outline diagram (**c**) shows the modern channels and lobes on the Tista megafan. The longitudinal profiles along the east, the center and the west of the megafan (Le, Lc, Lw) are shown in (**d**). Two transverse profiles (Tpf1 and Tpf2) across the Tista megafan are shown on (**e**). Note high vertical exaggeration due to low overall slope. (Following Chakraborty and Ghosh 2010)

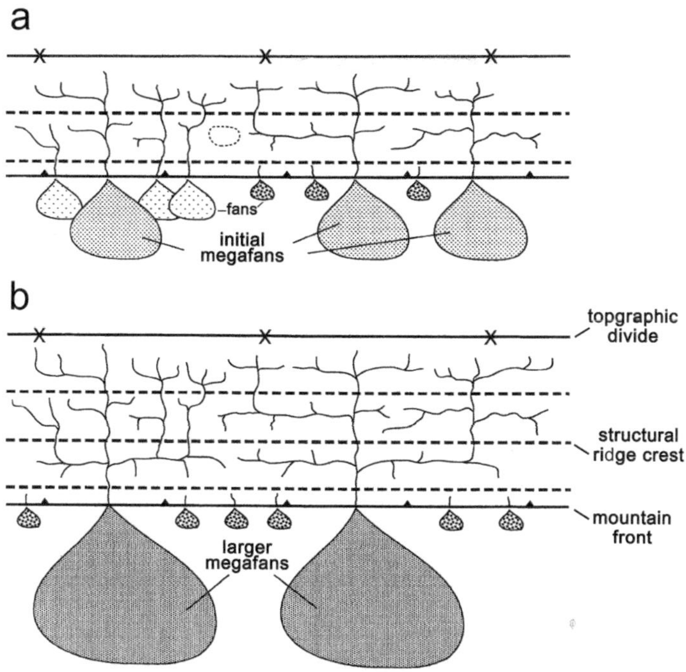

Fig. 2.4 Schematic model for evolution of megafans during propagation of a fold-thrust mountain front toward the foreland and the gradual (**a**–**b**) integration of the sourcing basins by headward capture resulting in well integrated drainage networks feeding a few megafans with alluvial fans inbetween. (after Horton and DeCelless 2001)

as part of the developing mountain front (Fig. 2.4). Well-known megafans are the Kosi and the Tista in India and the Bangladesh Himalayan foreland basin (Chakraborty and Ghosh 2010; Fig. 2.3a), the Rio Parapeti, Rio Pilcomayo, Rio Bermejo, Rio Salado (Horton and DeCelles 2001), and the Taquari megafan in front of the Andes, as well as the Okavango megafan of Botswana (Stanistreet and McCarthy 1993).

2.3.1 The Kosi and the Okavango Megafans

The Kosi River originates in Tibet; flows through the Nepal Himalaya, surrounding Mount Everest; and emerges in India, debouching into the Gangetic plains (Chakraborty and Ghosh 2010). It drains an area of 59,000 km^2 where annual rainfall ranges from 250 to 3560 mm. The Kosi River conveys an average annual runoff of 52×10^9 m^3, and its discharge is characteristically more continuous and less flashy. The concentration of its suspended load is usually 2–5 g/l but increases during the monsoon period to about 18 g/l. The mean annual sediment load of the Kosi is about 130×10^6 m^3.

The *Kosi megafan* is located in north Bihar, India, and its apex is in Nepal. The megafan covers an area of about 11,000 km^2. The overall longitudinal profile of the

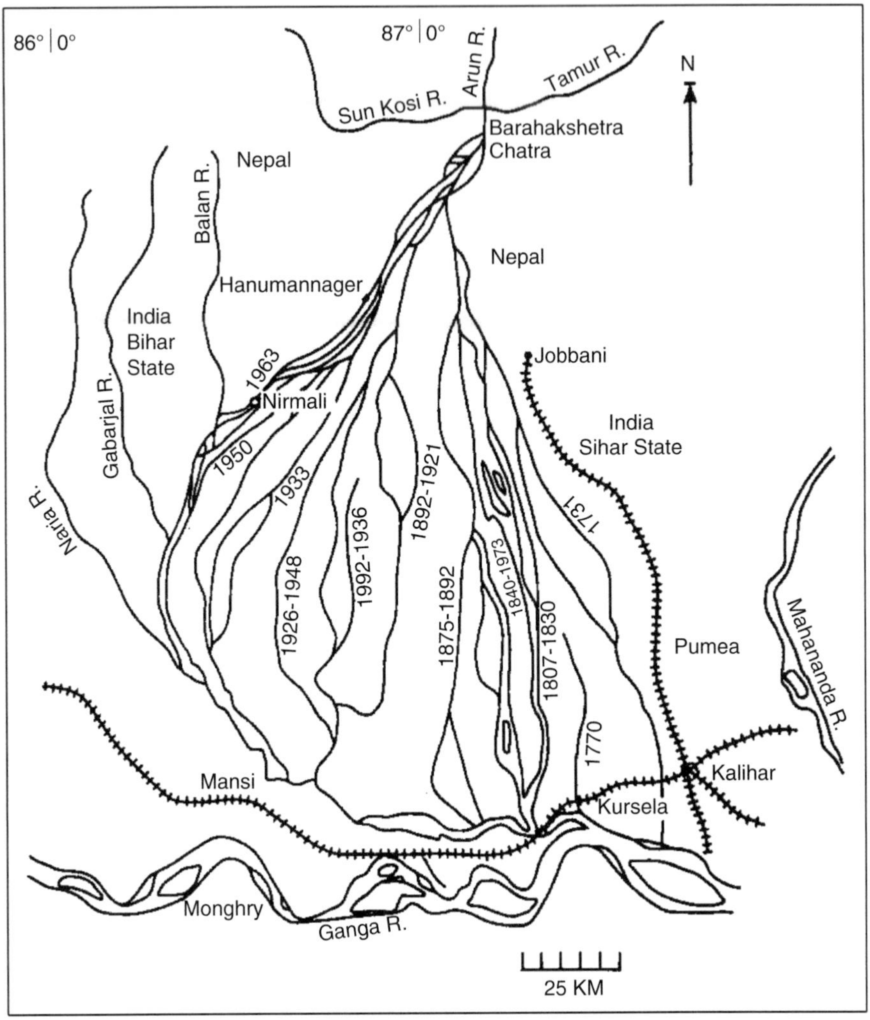

Fig. 2.5 The megafan of the Kosi River, India. Dated is its migration westward: 113 km in 228 years in progressive and episodic jumps. (Following Parker et al. 1998)

Kosi River floor is concave-upward, with slopes up to 0.0005 low (Fig. 2.2) and vegetation at its distal channels. Parts of the Kosi surface show four morphological levels, up to about 3 m above the active bed, representing stages in downcutting. The Kosi main feeder underwent rapid westward lateral migration to its present position at the western extremity of the megafan (Fig. 2.5). Westward tilting of the fan seems unlikely because its east side is now lower than the area to the west (Wells and Dorr 1987). The Kosi migrated 113 km in 228 years in progressive and episodic jumps of up to 19 km/year (Wood 1976; Bridge 2003; Chakraborty et al. 2010), thus sweeping across almost the whole surface of its fan at an average rate of 0.5 km/year, causing great damage to life and property (Gole and Chitale 1966).

The Kosi channel system changes from a proximal braided outline to a distal meandering pattern with an increasing sinuosity index of 1.4 - 1.6. The channels provide a continuous downstream flow of water that often terminates in swampland regions. The Kosi fan surface has an appearance of a flat, monotonous alluvial plain with old abandoned channel courses, lakes and oxbow lakes, swamps, bars, islands a few kms in length, irrigating canals, migrating sand mounds up to 2 m height, levees, and cutoff abandoned channels that continue to carry significant discharge contributed by the high groundwater table. Ground water seepage feeds perennial channels that flow over the megafan and are in places incised up to 6 m. Interchannel areas constitute the largest morphological units. Deeper parts in the megafan have been filled, and some have been gradually abandoned. Some of the low-lying grounds remain water logged for most of the year.

The Okavango River (McCarthy et al. 1992; Stanistreet and McCarthy 1993) receives water and sediments from the Central Highlands of Angola, where it is named the Cubango River that drains a catchment of 180,000 km^2 and terminates in a broad flat fault-bounded depression in northern Botswana. *The Okavango megafan* spreads over about 18,000 km^2 and is 150 km in axial length. Its average discharge is approximately 335 m^3/s with seasonal peaks up to 700 m^3/s supporting vegetation in permanent and seasonal swamps.

The Okavango records the lowest gradient of any previously described subaerial fan system – 0.00036 (Fig. 2.2). It includes newly activated distributary channels and re-flooded old oxbows lakes, islands, seasonal and permanent wetlands, and swamps transected by channels and confined by densely vegetated peat banks. In the proximal reaches, the river is sandy and meandering, anastomosing in the medial zone and continuing to a distal, poorly channelized and unconfined zone. In the spectrum of fans, the Okavango provides the end member as a meandering, low sinuosity "losimean" case, dominated by anastomosing meanders and a low sinuosity belt with highly vegetated, interchannel swamps forming thick peats.

References

Al-Farraj A, Harvey AM (2005) Morphometry and depositional style of late Pleistocene alluvial fans: Wadi Al-Bih, northern UAE and Oman. In: Harvey AM, Mather AE, Stokes M (eds) Alluvial fans, geomorphology, sedimentology, dynamics. The Geological Society, London

Blair TC, McPherson JG (1994) Alluvial fans and their natural distinction from rivers based on morphology, hydraulic processes, sedimentary processes and facies assemblages. J Sediment Res 64:450–489

Bridge JS (2003) Rivers and floodplains: forms, processes and sedimentary record. Blackwell Science, Oxford

Chakraborty T, Ghosh P (2010) The geomorphology and sedimentology of the Tista megafan, Darjeeling Himalaya: implications for megafan building process. Geomorphology 115:252–266

Chakraborty T, Kar R, Ghosh P, Basu S (2010) Kosi megafan: historical records, geomorphology and the recent avulsion of the Kosi River. Quat Int 227:143–160

Dec T (1992) Textural characteristics and interpretation of second-cycle, debris-flow- dominated alluvial fans (Devonian of Northern Scotland). Sediment Geol 77:269–296
DeCelles PG, Gray MB, Ridgway KD, Cole RB, Srivastava P, Pequera N, Pivnik DA (1991) Kinematic history of a foreland uplift from Paleocene synorogenic conglomerate, Beartooth Range, Wyoming and Montana. Geol Soc Am Bull 103:1458–1475
Gole CB, Chitale SV (1966) Inland delta building activity of Kosi River. J Hydraul Div Am Soc Civil Eng HY-2:111–126
Gómez-Villar A, Álvarez-Martínez J, García-Ruiz JM (2006) Factors influencing the presence or absence of tributary-junction fans in the Iberian Range, Spain. Geomorphology 81(3):252–264
Gupta S (1997) Foreland basin Himalayan drainage patterns and the origin of fluvial megafans in the Ganges. Geology 25:11–14
Harvey AM (1984) Debris flows and fluvial deposits in Spanish quaternary alluvial fans: implication for fan morphology. In: Koster EH, Steel R (eds) Sedimentology of gravel and conglomerates. Canadian Society of Petroleum Geologists, Calgary, mem 10:123–132
Horton BK, DeCelles PG (2001) Modern and ancient fluvial megafans in the foreland basin system of the central Andes, southern Bolivia: implications for drainage network evolution in fold-thrust belts. Basin Res 13:43–63
Leier AL, DeCelles PG, Pelletier JD (2005) Mountains, monsoons and megafans. Geology 33:289–292
Mather AE, Harvey AM, Stokes M (2000) Quantifying long-term catchment changes of alluvial fan systems. Geol Soc Am Bull 112(12):1825–1833
McCarthy TS, Ellery WN, Stanistreet IG (1992) Avulsion mechanisms on the Okavango fan, Botswana: the control of a fluvial system by vegetation. Sedimentology 39:779–795
Mikesell LR, Weissmann GS, Karachewski JA (2010) Stream capture and piracy recorded by provenance in fluvial fan strata. Geomorphology 115:267–277
Muravchik M, Bilmes A, D'Elia L, Franzese JR (2014) Alluvial fan deposition along a rift depocentre border from the Neuquén Basin, Argentina. Sediment Geol 301:70–89
Parker G, Paola C, Whipple KX, Mohrig D (1998) Alluvial fans formed by channelized fluvial and sheet flow. I: theory. J Hydraul Eng 124(10):985–995
Rickenmann D, Schiedl C (2013) Debris flow runout and deposition on the fan. In: Schneuwly-Bollschweiler M, Stoffel M, Rudolf-Miklau F (eds) Dating torrential processes on fans and cones, methods and their application for hazard and risk assessment, Advances in Global Change Research, vol 47. Springer, Dordrecht
Stanistreet IG, McCarthy TS (1993) The Okavango fan and the classification of subaerial fan systems. Sediment Geol 85:115–133
Stock JD (2013) Waters divided: a history of alluvial fan research and a view of its future. In: Wohl E (ed) Treatise on geomorphology, Fluvial Geomorphology, vol 9. Academic, San Diego, pp 413–458
Stokes M, Mather AE (2015) Controls on modern tributary-junction alluvial fan occurrence and morphology: High Atlas Mountains, Morocco. Geomorphology 248:344–362
Sweeney M, Loope DB (2001) Holocene dune-sourced alluvial fans in the Nebraska Sand Hills. Geomorphology 38:31–46
Talbot MR, Williams MAJ (1978) Erosion of fixed dunes in the Sahel, Central Niger. Earth Surf Process 3:107–113
Wells N, Dorr JA (1987) A reconnaissance of sedimentation on the Kosi alluvial fan of India. In: Ethridge F, Flores RM, Harvey AM (eds) Recent development in fluvial sedimentology, vol 39. Special Publication, Society of Economic Paleontologist and Mineralogists, pp 51–61
Wells SG, Harvey AM (1987) Sedimentologic and geomorphic variations in storm-generated alluvial fans, Howgill Fells, Northwest England. Geol Soc Am Bull 98:182–198
Whipple KX, Trayler C (1996) Tectonic controls on fan size: the importance of spatially variable subsidence rates. Basin Res 8:351–366
Wood G (1976) The legacy of the past: the agrarian structure. In: Joy JL, Everitt E (eds) The Kosi symposium: institute of development studies. pp 93–112

Slope Gradients 3

Abstract

Discharge is the first-order factor controlling fan slope. The greater the stream discharge, the gentler and flatter will be the slope of the fan. The slope is inversely proportional to the size of the catchment area and to the fan area. Longitudinal fan profiles typically show a concave-upward outline. Concavity is the result of the down-fan water loss, the spread of the water, and the decrease in the caliber of the bed material.

Most fans reported in the literature are in the 1–6° slope range. These typical high slopes of the fans make the flow almost invariably supercritical (Fr > 1). Megafans represent an ultralow slope category, ranging up to 0.001 - 0.00023. Humid-region fans have lower slopes than fans of arid regions. Debris flows form steeper fan gradients compared to fluvially dominated fans and to intermediate flows.

Keywords

Concave-upward profile · Lateral channel diversion · Profile segmentation · Intersection point · Transport efficiency · Froude number

3.1 Slopes and Profiles

Longitudinal fan profiles typically show a slope decrease toward the tow, giving rise to a slight or strong *concave-upward* longitudinal profile (Figs. 3.1 and 3.2).

Concavity is the result of a down-fan decrease in water supply following a down-fan water loss and due to the spread of the water. The slope of the highest and oldest fan segment near the mountain front is the steepest site on fans, usually slightly steeper than the average values measured from the apex to the toe. A down-fan decrease in stream competence results in an increased sediment storage at the proximal parts of the fan with a gradual slope diminution from the apex through the

© Springer Nature B.V. 2019
D. Bowman, *Principles of Alluvial Fan Morphology*,
https://doi.org/10.1007/978-94-024-1558-2_3

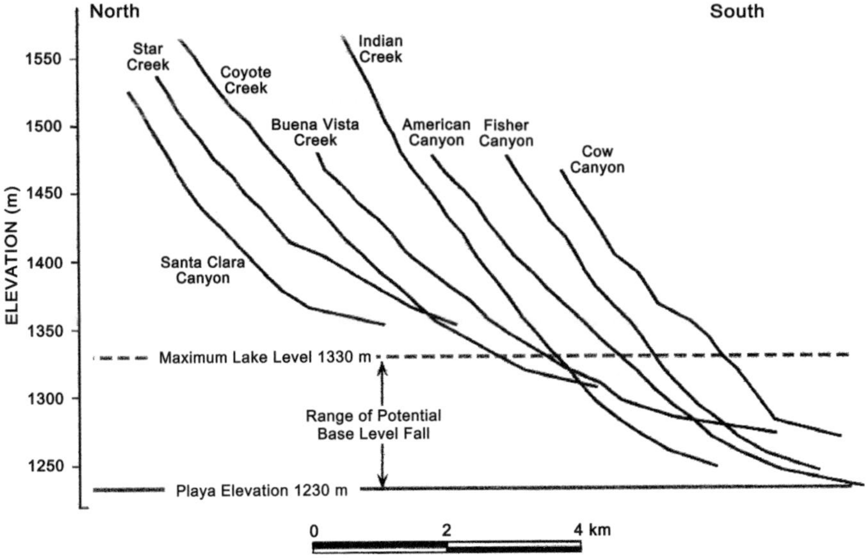

Fig. 3.1 Concave-shaped profiles of alluvial fans in the western edge of the Buena Vista Valley, Nevada. Part of the fans are related to the high stand of Pluvial Lake Lahontan and others to the present playa floor. (Redrawn from Ritter et al. 2000)

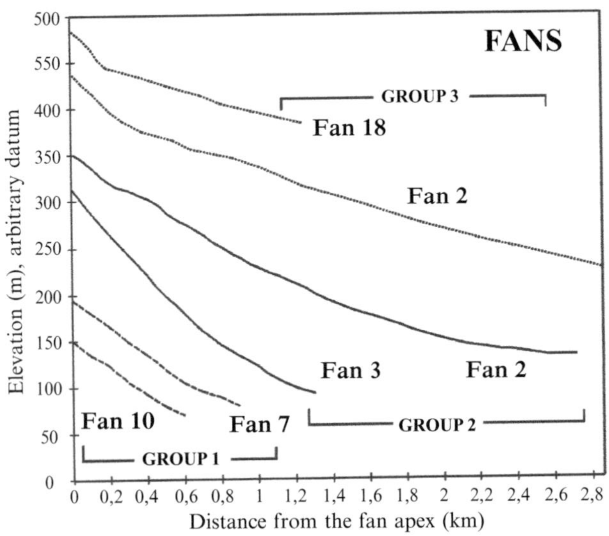

Fig. 3.2 Fan longitudinal profiles from the Betic Cordillera, Spain. Group 1 shows steep and almost constant slopes. Groups 2 and 3 show profiles with different concavities. The change from steeper to smoother segments occurs along the various profiles at different distal or proximal locations. Vertical exaggeration 5. (Redrawn from Viseras et al. 2003)

medial zone down to the toe. Higher proportions of mud and coarser sediments at the same discharge typically produce steeper fans. Alluvial fans tend to have higher mean fan slopes when they are dominated by debris-flow deposits rather than by fluvial sediments. Tectonic activity with dominance of mountain uplift over erosion will form steeper and smaller fans. Conversely, a long-term dominance of erosion rate over rock uplift will yield gentler fan slopes.

Fan slope is likewise linked to the sediment flux from the catchment. The slope is inversely dependent on the "efficiency" of the sediment transport system. High sediment and water discharge will form a large fan with a low surface slope. Fan slope is as well dictated by the balance between the flux of material removed from a catchment and the rate of accommodation generation. A continuous increase of an accommodation space favors a vertical accumulation and contributes to steeper slopes. A constant sinking of a basin bottom maintains a steeply sloping accumulation of sediments as opposed to expansion and progradation of fans in stable conditions which result in lower slopes.

Successively younger entrenched channels show lower gradients. The longitudinal profile of the upper fan surface and of the entrenched and active channels converges gradually and merges distally down-fan. Their profiles may, however, also display parallelism. When the upper zone of a fan is uplifted, an upslope divergence of the profiles of the fan surface and of its entrenched channels is expected.

Cross profiles of fans, parallel to the mountain front, show convexity which is greatest near the apex and becomes subdued toward the mid-fan and the distal areas (Fig. 1.7). Fans are rarely perfectly conically shaped but are often longer normal to the mountain front, which makes their medial radial line (Fig. 1.1) more gently sloped. Diversion of the flow from the medial radial line laterally toward the ever steeper flanks initiates lateral channel diversion, a basic process which leads to the formation and lateral growth of fans and to their subsequent merging.

Channels issuing from mountain fronts do not change their gradient significantly upon entering the fan apex (Bull 1964). In most of the alluvial fans described in literature, the slope of the feeder channel in the mountains coincides with or is slightly higher than the slope of the proximal fan surface (Bull 1962; French 1987), expressing the trend of smoothing following the deposition of fans at the mountain front.

With increasing distance from the apex down-fan, apart from decrease in water discharge and in the caliber of the bed material and in stream power, an overall decrease in the channels' width and depth can be observed. Down-fan the smallest flow rates and the lowest slope-gradient-combination is to be found (Cao et al. 2009). Higher threshold slopes required to entrain the coarser particles are found in the upper fan. Short and small fans with a more constant linear slope lack a significant textural fining down-fan. Longer fans which allow a gradually improved sorting enhance concavity. Steepening of gradients down-fan toward the distal zone may indicate that overriding factors are at work at the distal edge of the fan, as, for instance, a lowering base-level.

A profile segment is a fan section of uniform gradient that is bounded up- and down-fan by surfaces of different slope gradients. Segmented radial fan profiles are those with abrupt changes in slope. Breaks in slope can be located by drawing topographic profiles along fan radii (Hooke 1972), and by Total Station surveys may reveal the presence of fan segments. Segmentation may occur due to reoccurring tectonic uplift episodes of the mountainous drainage basin, causing deposition of fans, one superimposed on top of the former, with a steepening trend toward the highest and youngest fan (Bull 1964, 1977; Denny 1967; Hooke 1967). Segmented fan profiles can result as well from entrenchment events unrelated to any tectonic control. Such *profile segmentation* may ensue from high discharge flow events which incise the fan and form an entrenched smoother channel segment followed by deposition of a steeper lobe.

The highest gradients are usually to be found at the apex of fans. Channel downcutting is associated with smoothing of the gradient. Successive deeper entrenchments form gradually smoother slope gradients. A succession of longitudinal profiles converges down-fan. Fan toe entrenchment may too contribute to segmentation by headward entrenchment. Segmentation may result from knickpoints along channels entrenched through indurated soils. *Intersection point*, a term introduced by Hooke (1967), is the site where a relatively low-gradient entrenched channel emerges onto the steeper fan surface and forms a break in slope (Fig. 3.3).

Longitudinal slopes from apex to toe of most fans reported in the literature are in the 1–6° slope range, attaining a wider 1–12° slope range for shorter-radius fans (Saito and Oguchi 2005; Stanistreet and McCarthy 1993). Very low fan gradients of <1° have also been reported. Debris-flow-dominated fans are typically steeper than fluvially dominated fans which tend to be larger and longer. Megafans represent the ultralow slope category, ranging 0.001–0.00023 (<2′). Globally, the slopes on fans are continuous to slopes of rivers without any specific gap (Saito and Oguchi 2005). Alluvial fans can be approached as a *steep fluvial-dominated environment* which provides the opportunity to focus on the control of *steepness* and *radiating unconfined flows* over fluvial processes. Bedload transport equations have not been verified

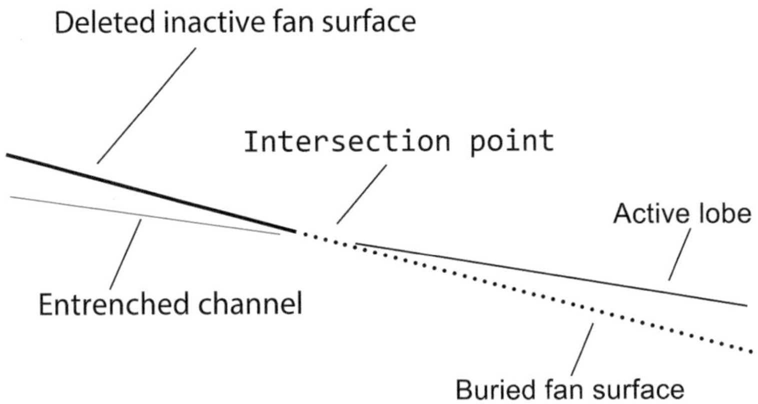

Fig. 3.3 The intersection point

for use on steep alluvial fans with supercritical flow conditions that likely occur on many arid-region fans with substantial sand fractions on their bed (Stock 2013).

3.2 Controlling Factors

Discharge Flume experiments (Whipple et al. 1998; Milana and Tietze 2002), mathematical models (Parker et al. 1998), and field studies (Bull 1964) all show that discharge is the first-order factor controlling fan slope. Fan slope is inversely proportional to discharge and is dependent in a nonlinear way both on discharge and on the sediment size. As long as shear stresses are well above critical, water discharge exerts a much stronger control on fan morphology than grain size (Parker et al. 1998). Drainage basin area combined with mean annual rainfall control the slope through the *transport efficiency* (Milana and Ruzycki 1999), a property that describes how fast the system removes sediment by an amount of the water traversing the fan. Efficiency is related to the mean annual runoff and to the maximum power peaks, approximated by the peak flood. Two fans with equal drainage basin areas but under different climate regimes and dissimilar runoffs will display different slopes. Transport efficiency on fans needs steeper slopes when the hydrological regime is low. When the gradient of the channel is insufficient to transport the sediment, steepening will take place due to sediment aggradation. The greater the stream discharge, the gentler and flatter will be the slope of the fan. The smaller the drainage basins, the runoff, and the peak flow, the steeper the fans will become. Humid-region fans have lower slopes than arid-region fans of a similar drainage basin area (Fig. 3.4).

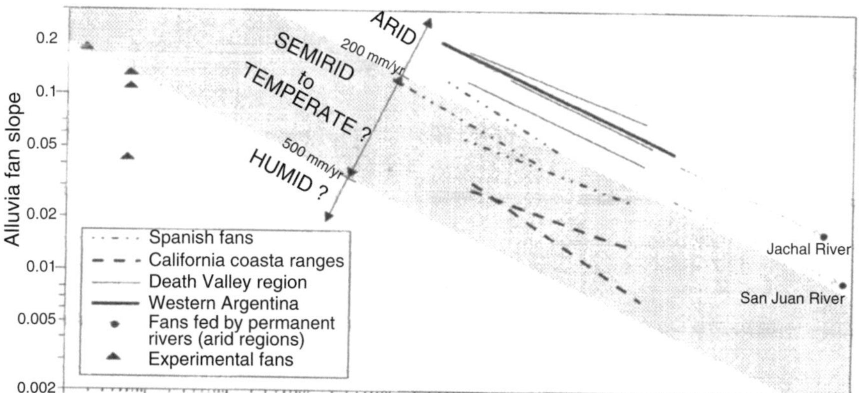

Fig. 3.4 Regression lines for the relation between the drainage basin area and the slope of alluvial fans in different climatic regions. A tentative separation of the arid region from the humid region is proposed. Humid regions with increased rainfall and transport efficiency have lower slopes than arid-region fans of similar drainage basin area. (From Milana and Ruzycki 1999)

Basin Area/Fan Slope Relation Fan slope is inversely proportional to the size of the catchment and the fan area (Fig. 3.5a, b). The negative relation of fan gradient to drainage area (Fig. 3.6) is expressed by

$$S = aA^{-b} \tag{3.1}$$

(Harvey 1989, 1990; Milana and Ruzycki 1999) where S is the fan gradient, A is the drainage area, and a and b are constants related to the regression line. Slopes of fans sourced by a drainage basin which underwent recent changes are expected to show a lower correlation index. Fan slopes correlate positively well with the basin relief ratio, i.e., the basin height divided by basin length (Schumm 1956, Fig. 3.5c). Melton (1965) suggested the ruggedness of the drainage basin as a controlling factor which is positively related to the slope of the upper fan surface. The suggested functional relationship is

$$S = a\left(H / \sqrt{A}\right)^{n} \tag{3.2}$$

where H is the vertical relief of the drainage basin above the apex, A is the drainage basin area, and a and n are an undetermined coefficient and an exponent, respectively.

Grain Size Sediment supply-quantity and grain size are important controls of fan gradients. The resistance of the catchment lithology controls the production rate and the size of debris supplied to the fans. Rock resistance, if high, shows by coarser sediments, steeper upper fan segments, and smaller fans. Debris flows form steeper fan gradients compared to water laid fine-grained, fluvially dominated fans.

The progressive decay of slope down-fan is related to the reduction of the coarse bedload with distance, shown by the decrease in the maximum and mean particle size and decrease in the transport rate down-fan (Stock et al. 2007; Fig. 3.7). However, over large fractions of the channel length, the grain size reduction cannot fully explain the channel slope reduction. To assess the role of grain size fining and the decline in bedload flux on fan long-profile slopes, we should collect the grains for distribution analysis at the same spot that the average flow depth and slope are collected (Stock 2013).

Slopes are adjusted to transport the supplied sediment load: higher threshold slopes are required to entrain the coarser particles in the upper fan, whereas lower slopes are enough to entrain the fine grains down-fan. Decrease in grain size which affects the decay in slope cannot occur in a well-sorted sedimentary environment. Good sorting already on leaving the source area will cause a low rate of grain size reduction, and no down-fan concavity is expected to develop which may result in a constant fan slope. Short fans which do not allow significant sorting tend toward a constant slope.

Hydraulics The steepness of the slope of alluvial fans significantly affects the flow regime and the shear stress (Blair and Mcpherson 1994). The effect can be illustrated

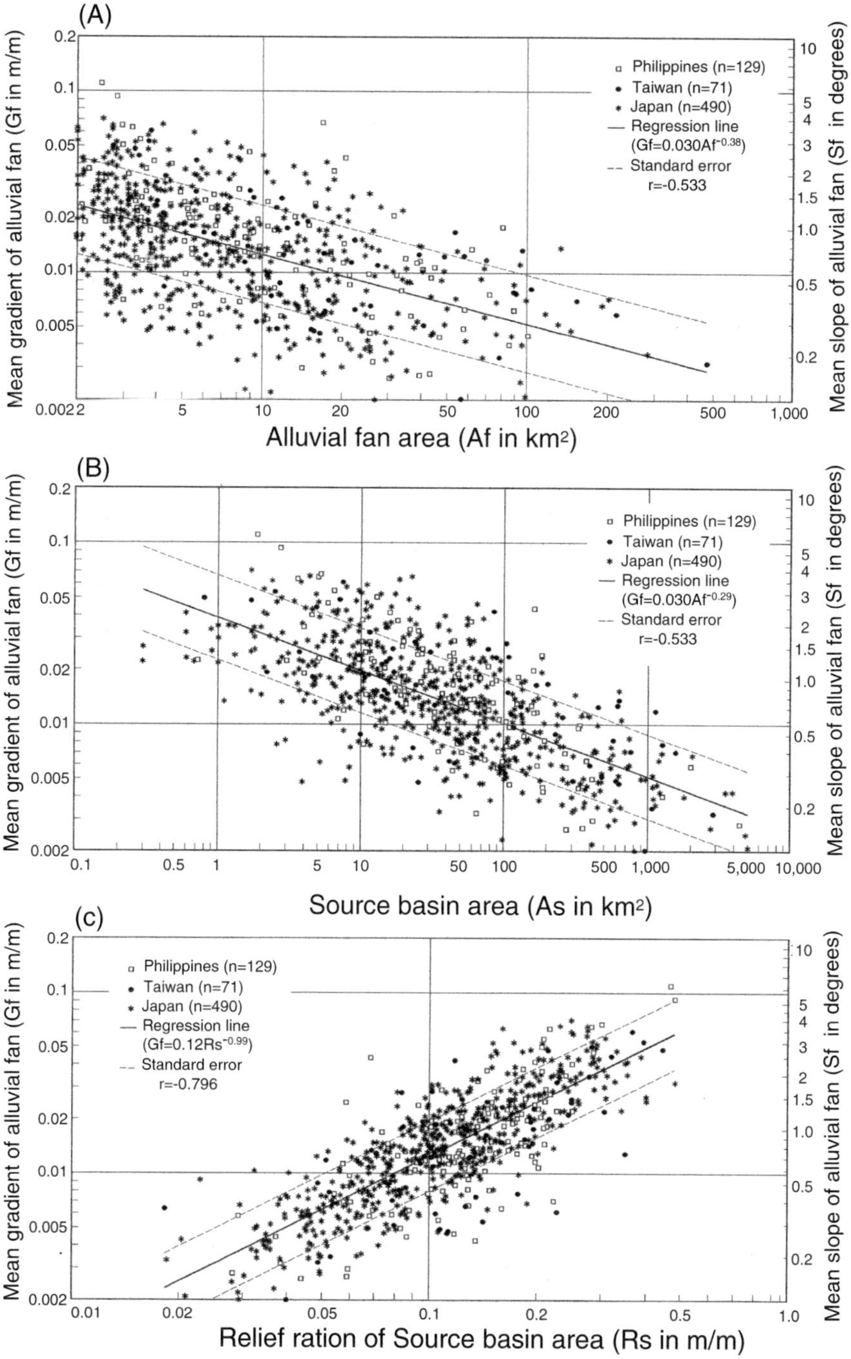

Fig. 3.5 (**a**) Logarithmic mean fan gradient versus logarithmic fan area, showing a negative correlation. (**b**) Logarithmic mean fan gradient versus logarithmic source basin area, showing a negative correlation. (**c**) Logarithmic mean fan gradient versus logarithmic basin relief ratio (following Schumm 1956), showing a positive correlation. Data from Japan, Taiwan and the Philippines. (From Saito and Oguchi 2005)

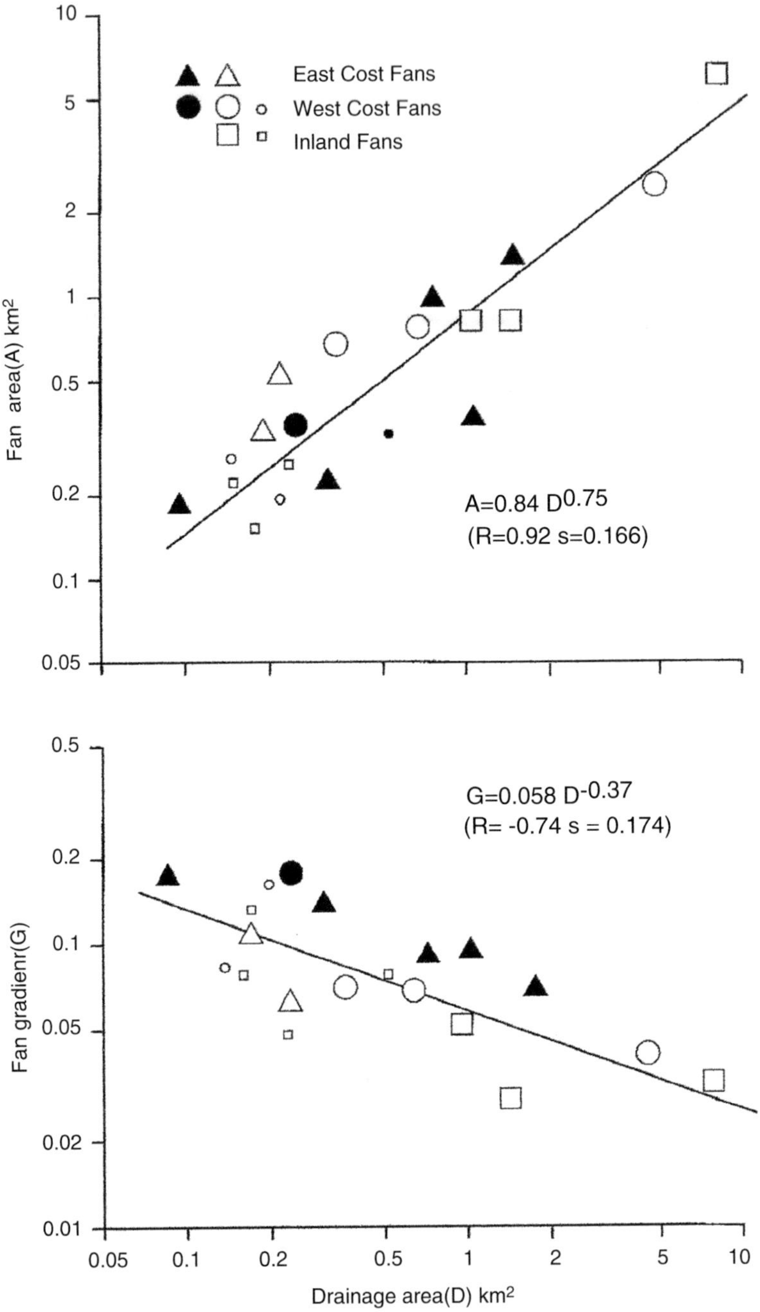

Fig. 3.6 Response of fan area and gradient to drainage area shown by regression plots. Included are dissected fans (solid symbols), aggrading fans (large open symbols), and confined fans (small open symbols, following Harvey et al. 1999)

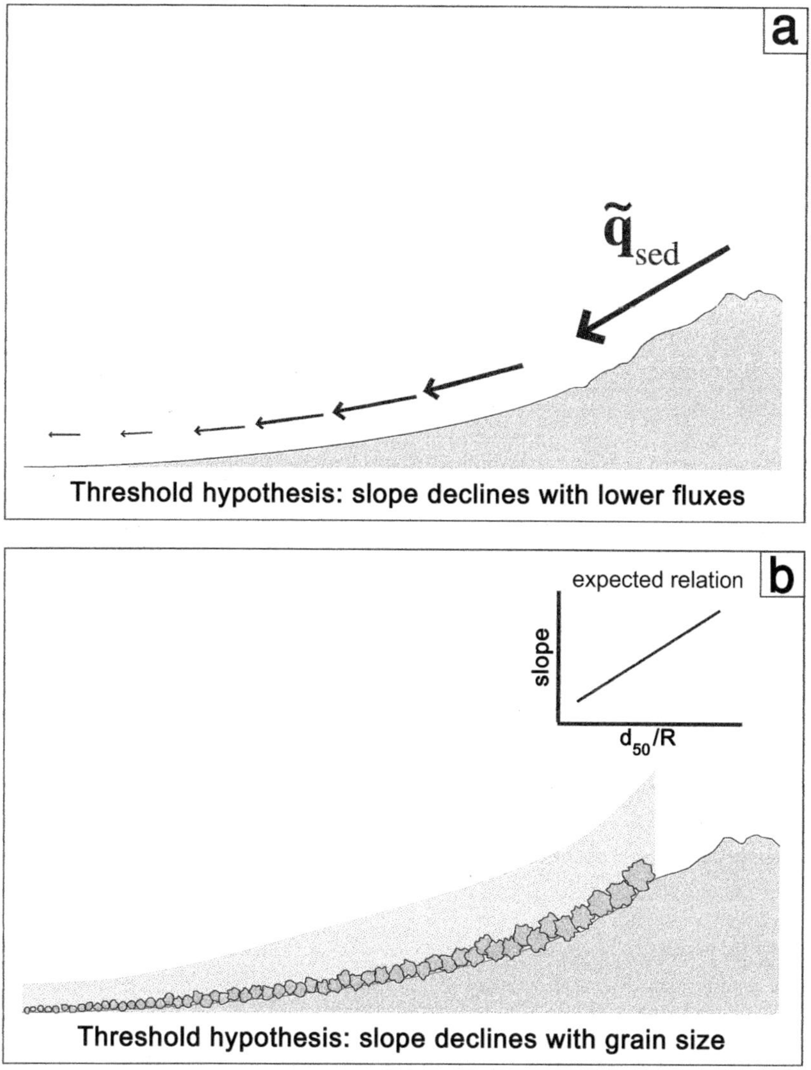

Fig. 3.7 Factors controlling the alluvial fan slope: Declining transport rate $\mathbf{q}_{sed}$ (**a**) and declining grain size down-fan (**b**) lead to slope reduction. Inset shows decrease of channel slope with decrease of the ratio grain size to depth d_{50}/R. (Modified after Stock et al. 2007). The slope decreases as gravel supply rate is reduced. Strong covariance is expected between maximum clast size and local surface slope. The coarsest grain fraction that can be entrained decreases down-fan with the fan slopes decline

by the Manning equation (Barnes 1967) where velocity is proportional to the square root of the slope

$$V = D^{2/3} S^{1/2} / n \tag{3.3}$$

where V is the average flow velocity in m/s, S is the slope in m/m, D stands for flow depth in meters, and n is the Manning's Roughness Coefficient. The effect is further illustrated by the Froude number

$$\mathrm{Fr} = V / g^{1/2} D^{1/2} \tag{3.4}$$

where V is the average flow velocity in m/s, g is the gravitational acceleration constant 9.8 m/s^2, and D is flow depth in meters. The typical high slopes of fans and the relative high flow velocities and shallow flow depth make the flow on fans almost invariably supercritical which is a fast and shallow flow with pulsations, defined by the dimensionless Froude number greater than one (Fr > 1). Such flow has a relative high energy state which affects the potential for entrainment and erosion. The greater the Froude number, the greater the irregular fluctuations of the flow speed, the formation of eddies, the mixing, and the chaotic changes in direction.

Alluvial fans present a steep alluvial environment with flow velocities that may range very high, 3–9 m/s with estimates up to 15 m/s. The hydraulic radius, i.e., the channelized cross-sectional area relative to its wet perimeter, shows a rapid, often exponential, decline down-fan (Stock et al. 2007; Fig. 3.8), where following flow expansion the bankfull depths grow shallower and the depositional trend takes over.

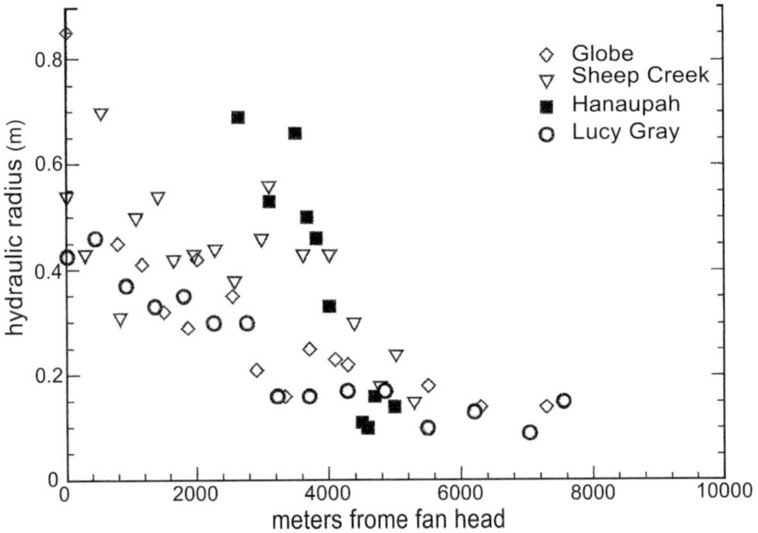

Fig. 3.8 The typically down-fan decline in the hydraulic radius from 0.5–0.7 m at fan heads to 0.1–0.2 m at the distal fan. Transport capacity declines exponentially down-fan as the radius decreases. Based on data from four fans of the western USA: Globe, Sheep Creek, Hanaupah, and Lucy Gray. (Following Stock et al. 2007)

References

Barnes HH Jr (1967) Roughness characteristics of natural channels. US Geol Surv Water supply paper 1849, p 213

Blair TC, McPherson JG (1994) Alluvial fans and their natural distinction from rivers based on morphology, hydraulic processes, sedimentary processes, and facies assemblages. J Sediment Res 64A:450–489

Bull WB (1962) Relation of textural (CM) patterns to depositional environment of alluvial fan deposits. J Sediment Petrol 32:211

Bull WB (1964) Geomorphology of segmented alluvial fans in western Fresno County, California. US Prof pap 352-E: 89–129

Bull WB (1977) The alluvial fan environment. Prog Phys Geogr 1:222–270

Cao L, Zhang K, Zhang W (2009) Detachment of road surface soil by flowing water. Catena 76:155–162

Denny CS (1967) Fans and pediments. Am J Sci 265:81–105

French RH (1987) Hydraulic processes on alluvial fans. Development in Water Science 31

Harvey AM (1989) The occurrence and role of arid-region alluvial fans. In: Thomas D (ed) Arid region geomorphology. Belhaven/Frances Printer Publ, London

Harvey AM (1990) Factors influencing quaternary alluvial fan development in Southeast Spain. In: Rachocki A, Church M (eds) Alluvial fans: a field approach. Wiley, Chichester, pp 247–269

Harvey AM, Silva PG, Mather AE, Goy JL, Stokes M, Zazo C (1999) The impact of Quaternary sea-level and climatic change on coastal alluvial fans in the Cabo de Gata ranges, southeast Spain. Geomorphology 28(1–2):1–22

Hooke RL (1967) Processes on arid-region alluvial fans. J Geol 75:438–460

Hooke RL (1972) Wisconsin and Holocene tectonic geomorphic evidence for late deformation, Death Valley, California. Geol Soc Am Bull 83:2073–2098

Melton MA (1965) The geomorphic and the palaeoclimatic significance of alluvial deposits in southern Arizona. J Geol 66:1–38

Milana JP, Ruzycki L (1999) Alluvial-fan slope as a function of sediment transport efficiency. J Sediment Res 69:553–562

Milana JP, Tietze KW (2002) Three-dimensional analogue modelling of an alluvial basin margin affected by hydrological cycles: processes and resulting depositional sequences. Basin Res 14:237–264

Parker G, Paola C, Whipple KX, Mohrig DC (1998) Alluvial fans formed by channelized fluvial and sheet flow. I: theory. J Hydraul Eng 124(10):985–995

Ritter JB, Miller JR, Husek-Wulfo J (2000) Environmental controls on the evolution of alluvial fans in BuenaVista Valley, North Central Nevada, during late Quaternary time. Geomorphology 36:63–87

Saito K, Oguchi T (2005) Slope of alluvial fans in humid regions of Japan, Taiwan and the Philippines. Geomorphology 70:147–162

Schumm SA (1956) Evolution of drainage systems and slopes in badlands at Perth Amboy, New Jersey. Geol Soc Am Bull 67:597–546

Stanistreet IG, McCarthy TS (1993) The Okavango fan and the classification of subaerial fan systems. Sediment Geol 85:115–133

Stock JD (2013) Waters divided: a history of alluvial fan research and a view of its future. In: Wohl E (ed) Treatise on geomorphology, vol 9, Fluvial Geomorphology. Academic, San Diego, pp 413–458

Stock JD, Schmidt KM, Miller DM (2007) Controls on alluvial fan long-profiles. Geol Soc Am Bull 120:619–640

Viseras C, Calvache ML, Soria JM, Fernandez J (2003) Differential features of alluvial fans controlled by tectonic or eustatic accommodation space. Examples from the Betic Cordillera, Spain. Geomorphology 50:181–202

Whipple KX, Parker G, Paula C, Mohring D (1998) Channel dynamics, sediment transport and the slope of alluvial fans: experimental study. J Geol 106:677–693

Fan Morphometry

4

Abstract

Subsidence or uplift of the accommodation space controls the morphometry of fans. The outlet spacings are often strikingly similar. The size of the source area is a major factor determining the fan area and its slope. Fan size is represented by the planimetric fan area (Af) that is applied as a proxy for sediment yield. Alluvial fans tend to show a strong log-linear positive relationship between the fan area (Af) and the plan-view drainage basin area (Ad). Larger drainage basins contribute proportionally less sediment to the growth of the fan area. Inconsistency between the fan area and its catchment area may stem from piracy, nourishment by pre-existing deposits or from karstic conditions. Drainage area versus fan slope shows in most cases a significantly high negative relationship. The terminology used to characterize the geometry of alluvial fans and their sourcing catchments is provided.

Keywords

Geometric characteristics · Morphometric ratios · Accommodation space · Outlet spacing · Trap efficiency · Sediment storage

4.1 Definitions

Research established geometric characteristics of fans and of their sourcing drainage basins and defined morphometric relations. The following terminology is used to characterize alluvial fans:

Fan outlet spacing – the distance between fan outlets along a mountain front (Fig. 4.1).

Fan area – the planimetric area of each fan.

© Springer Nature B.V. 2019
D. Bowman, *Principles of Alluvial Fan Morphology*,
https://doi.org/10.1007/978-94-024-1558-2_4

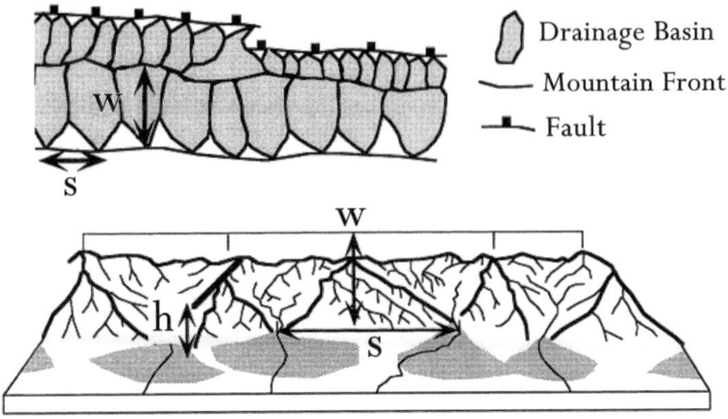

Fig. 4.1 Drainage basins, outlets, and alluvial fans along a faulted linear mountain front. Indicated is the topographic half width (w) of the ridge measured from the mountain front to drainage divide, the outlet spacing (s) and the topographic relief (h) measured as the elevation difference between the base of the mountain front and the drainage divide. (From Talling et al. 1997)

Topographic/hydrographic apex – the highest point on the fan at the mountain front where flow is still confined before spreading out (Fig. 1.8). Under such conditions it also functions as a hydrographic apex. Due to entrenchment the spreading site migrates down-fan as an *intersection point.*

Fan radius – the greatest radial apex-toe distance that reflects the fan length (Fig. 1.1).

Longitudinal radial slope – the detailed or averaged slope along the active channel.

Fan width – maximum width normal to the *fan radius.*

Fan elongation – the width/radius ratio that expresses the general shape of the fan.

Sweep angle/flow expansion angle – the angle between the two widest separate channels on the fan (Fig. 1.1).

Fan relief – the vertical distance between elevated and the nearby lowest sites.

Fan maximum entrenchment – the difference in elevation between the highest fan surface at the apex and the active channel bed below it.

Incised channel percentage – the percentage of the fan radius occupied by incised channels representing the fan surface.

The following are the morphometric parameters of sourcing drainage basins:

Drainage basin area – the total planimetric area of the drainage basin.

Drainage basin length – the longest distance of the feeder channel from the divide to the fan outlet.

Drainage basin width – the distance measured at the middle of the longitudinal axis of the drainage basin.

Drainage basin maximum relief – the difference in height between the highest point in the catchment area and the outlet.

Relief ratio of the source area – the basin maximum relief height divided by the basin length.
Basin ruggedness/relative relief number – the ratio of the drainage basin maximum relief to the square root of its area ($H_d/A_d^{0.5}$) (Melton 1965; Kostaschuck et al. 1986).
Feeder channel slope – the gradient of the main channel in the drainage area.

Accommodation space is the depositional basin area available for the accumulation of the fan sediments delivered from the drainage area. Accommodation space is provided by tectonic vertical movements including thrust activity (Fig. 1.4). The shape of the fan, its size, and spatial distribution are controlled by the accommodation space, its geometry, and the subsidence rate (Weissmann et al. 2005; Whipple and Trayler 1996). The fan length is the function of the ratio of volumetric sediment supply to the subsidence rate (Parker et al. 1998). This hypothesis should, however, be further tested in field sites of variable subsidence rates (Stock 2013).

High subsidence rates afford rapid accommodation space. Fans that build into deep grabens with high rates of tectonic subsidence tend to have a smaller area with a smaller distal growth and display less correlation of the fan area to the drainage basin area. Where subsidence is small, fans prograde farther. Fans may be thickest near rising mountains or lense shaped with the largest thickness in the mid-fan zone.

Adjacent drainage basins often show regular *outlet spacing*. The ratio of the distance from the main drainage divide to the base of the mountain front to the outlet spacing, which in effect is the ratio of the length to the width of catchments (Fig. 4.1), is strikingly uniform, within the narrow range of 1.91–2.23, indicating a certain morphological homogeneity (Hovius 1996). Outlet spacing is likely to be determined during the early stages of the drainage evolution (Talling et al. 1997) and probably reflects similar interacting conditions between mountain belts and erosional processes. Regularity in the outlet spacing has significant implications for the regular fan deposition along sedimentary basins. The length of lateral facies variations, controlled by the outlet spacing, is of importance in the development of petroleum reservoir units.

4.2 Fan Area vs Drainage Area

The size of the source area is a major factor determining the fan area and its slope. The relationship between fan area and the catchment area has been discussed among others by Denny (1965), Hooke (1968), Whipple and Trayler (1996) and by Allen and Hovius (1998). We would expect a good correlation between the catchment area and the volume of the fan, including its buried portion, which is, however, usually not known nor calculable. Fan size is, therefore, represented by the planimetric fan area (A_f) that is applied as a proxy for sediment yield and does not consider the thickness of the fan which may differ, dependent on the structure of the mountain front and of the accommodation space (Giles 2010). The relationship between volume and planimetric area within groups of alluvial fans was investigated in non-subsiding tectonic settings (Giles 2010) and indicated that the proportional power-law relationship between fan volume and fan area is not significantly different from 1:1. Fan area and fan volume would be equivalent representations of a fan's size. Substituting area as a proxy for fan volume is best acceptable for fans bordering non-subsiding basins. Fan volume may be underestimated in subsiding areas because of the increased buildup of thick, stacked fans without concomitant growth of the area. Bull (1962, 1964) showed that inspite of the volume/area problem, power-law relation exists between the fan area (A_f) and the feeding drainage basin area (A_d)

$$A_f = c\, A_d^{\,n} \tag{4.1}$$

where c and n are empirically derived coefficients, c is the graph intercept value, and n is the proportionality coefficient that represents the slope of a best-fit line. Alluvial fans tend to show a strong log-linear positive relationship between the fan area and the area of their associated catchment (Fig. 3.6). High correlations (0.84–0.98) have been reported for modern fans in many field sites (Bull 1962; Harvey 1989; Allen and Hovius 1998). The overall relationship between the fan area and the drainage basin area may, however, be invalidated by differences in the substrate and in rates of erosion. Fans derived from drainage basins characterized mainly by mudstone and shale are larger than fans derived from drainage basins characterized by sandstones (Bull 1964).The most frequent values of exponent n are usually within the range of 0.6–1.1, most of them are less than unity, reflecting storage in the drainage basin (Hooke 1968). Some cases of n = 0.3–0.5 and of n = 1.0–2.8 have also been reported, indicating an unusually small or large increase in the fan area.

Most fans described in the literature show a *basin area/fan area* – A_f/A_d – *ratio* in the range of 1–10. Extreme ratios such as 1–20 have also been reported (Allen and Hovius 1998). Dade and Verdeyen (2007) reported an A_f/A_d ratio range between 0.03 and 1.4. The implication of $n < 1.0$ is that larger drainage basins contribute proportionally less sediment to the growth of the fan area. The reason fans grow more slowly as drainage basins increase in size is that the additional sediment yields relatively decrease as basin size increases, presenting an internal feedback mechanism, which dampens the fan growth. Large basins are less likely to be covered by single precipitation events, and only a portion of the drainage basin will often actively contribute sediments to the fan. In addition, large drainage basins tend to have smaller slopes, which results in a smaller transport capacity. Diminishing additional sediment yields are expected in larger catchments also following greater run-off losses. Differences in *sediment storage* are also important: whereas in mountainous areas small catchments have narrow valley floors, unsuitable for sediment storage, wider catchment valleys offer an increased storage on the slopes and along the valley (Hooke 1968).

The intercept value c expresses the magnitude of the fan which is related to 1 km^2 drainage basin. C shows a wide range, mainly in the 0.1–1.0 reach with less frequent magnitudes of 1.0–2.6. The wide range of c values stems from the following reasons: different tectonic subsidence rates which control the availability of the depositional space; different erodibilities of the drainage basins; and variable catchment storage capabilities in the sourcing drainage basins. Diverse magnitudes of rainfall intensity and different fan ages may contribute as well to the wide range of c values. Harder lithologies as quartzite produce coarse but less debris and form by far smaller fans.

Fan area is controlled as well by its *trap efficiency*, defined by the fraction of the total sediment yield deposited on the fan (Whipple and Trayler 1996). Decrease in trap efficiency is the result of transport of sediment across and beyond the fan due to bypassing. Small fans relative to their catchment area may also form owing to lateral confinement or as a result of distal erosion by an axial river. An opposing trend of formation of extended fans with respect to the drainage basin can stem from huge landslide events (Crosta and Frattini 2004) or from postglacial transportation of large amounts of pre-existing paraglacial deposits, not eroded from the presently active drainage area.

The fan area/catchment area ratio may reflect as well evolution of the catchment area following piracy and capture which change the watershed boundary and magnitude. Partly captured basins may show an abnormally high fan area/basin area relationship. Inconsistency between fan area and its catchment area may also result from restrictions on the fan geometry in the accommodation space or from extensive karstic conditions with unique relationships between rainfall characteristics, surface runoff, and subsurface drainage. Paleosourcing areas of fans can be reconstructed based on preserved paleo-fan remnants while applying fan morphometric equations (Mather et al. 2000).

4.3 Fan Slope vs Drainage Area

The correlations between the catchment area (A_d) and the average fan slope (G) can be described by the power function

$$G = aA_d^{-b} \tag{4.2}$$

where a is a constant and b the exponent. Drainage area versus fan slope shows in most cases a significantly high negative relationship of 0.8–0.9 (Bull 1962; Silva et al. 1992; Harvey et al. 1999). The reason for the negative correlation is that increased water discharge results in higher flow velocity and bed shear stress which are capable of efficient transportation of sediments over lower slopes. Low-gradient fans are associated with large drainage basins. Alluvial fans tend to have higher mean fan slopes when dominated by a higher ratio of sediment to water or by debris-flow deposits (Harvey 1992). Drainage basin relief seems to have only a weak control on the fan area. Incorporating, however, basin relief together with basin area improves the correlation to fan area (Harvey 2005).

References

Allen PA, Hovius N (1998) Sediment supply from landslide-dominated catchments: implications for basin-margin fans. Basin Res 10:19–35

Bull WB (1962) Relations of alluvial fan size and slope to drainage basin size and lithology in Western Fresno County, California. United States Geological Survey, Prof Pap B 450:51–53

Bull WB (1964) Geomorphology of segmented alluvial fans in western Fresno County, California. US Geol Surv Prof Pap. 352E:89–12

Crosta GB, Frattini P (2004) Controls on modern alluvial fan processes in the central Alps, Northern Italy. Earth Surf Process Landf 29:267–293
Dade WB, Verdeyen ME (2007) Tectonic and climatic controls of alluvial-fan size and source-catchment relief. J Geol Soc Lond 164:353–358
Denny CS (1965) Alluvial fans in the Death Valley Region, California and Nevada. US Geol Surv, Washington, DC Prof. Pap. 466
Giles PT (2010) Investigating the use of alluvial fan volume to represent fan size in morphometric studies. Geomorphology 121:317–332
Harvey AM (1989) The occurrence and role of arid zone alluvial fans. In: DSG T (ed) Arid zone geomorphology. Belhaven, London
Harvey AM (1992) The influence of sedimentary style on the morphology and development of alluvial fans. Isr J Earth Sci 41:123–137
Harvey AM (2005) Differential effects of base level, tectonic setting and climatic change on Quaternary alluvial fans in the northern Great Basin, Nevada, USA. In: Harvey AM, Mather AE, Stokes M (eds) Alluvial fans, geomorphology, sedimentology, dynamics. The Geological Society London
Harvey AM, Silva PG, Mather AE, Goy JL, Stokes M, Zazo C (1999) The impact of Quaternary sea-level and climatic change on coastal alluvial fans in the Cabo de Gata ranges, Southeast Spain. Geomorphology 28:1–22
Hooke RL (1968) Steady-state relationships on arid-region alluvial fans in closed basins. Am J Sci 266:609–629
Hovius N (1996) Regular spacing of drainage outlets from linear mountain belts. Basin Res 8:29–44
Kostaschuck RA, Macdonald GM, Putnam PE (1986) Depositional processes and alluvial fan-drainage basin morphometric relationships near Banff, Alberta, Canada. Earth Surf Process Landf 11:471–484
Mather AE, Harvey AM, Stokes M (2000) Quantifying long-term catchment changes of alluvial fan systems. Geol Soc Am Bull 112(12):1825–1833
Melton MA (1965) The geomorphic and paleoclimatic significance of alluvial deposits in Southern Arizona. J Geol 73:1–38
Parker G, Paola C, Whipple KX, Mohrig D (1998) Alluvial fans formed by channelized fluvial and sheet flow I. Theory: J Hydraul Eng 124:985–995
Silva PG, Harvey AM, Zazo C, Goy JL (1992) Geomorphology, depositional style and morphometric relationships of Quaternary alluvial fans in the Guadalentin Depression (Murcia, southeast Spain). Z Geomorphol Neue Folge 36:325–341
Stock JD (2013) Waters divided: a history of alluvial fan research and a view of its future. In: Wohl E (ed) Treatise on geomorphology, vol 9., Fluvial Geomorphology. Academic, San Diego, pp 413–458
Talling PJ, Stewart MD, Stark CP, Gupta S, Vincent SJ (1997) Regular spacing of drainage outlets from linear fault blocks. Basin Res 9:275–302
Weissmann GS, Bennett GL, Lansdale AL (2005) Factors controlling sequence development on Quaternary fluvial fans, San Joaquin Basin, California, USA. In: Harvey AM, Mather AE, Stokes M (eds) Alluvial fans, geomorphology, sedimentology, dynamics. The Geological Society London
Whipple KX, Trayler CR (1996) Tectonic control of fan size: the importance of spatially variable subsidence rates. Basin Res 8:351–366

Drainage

5

Abstract

Alluvial fans are differentiated into the proximal, the medial, and the distal zone. Drainage processes operate on fans either along incised channels or in a sheetlike manner. The distributary systems change from a single-channel at the fan head to a braided-channel pattern down-fan. The channel size, the flow depth, and the velocity diminish down-fan where the channels become shallower, the discharge decreases, and the transport capacity is reduced.

The active fan area is at any time small and occupies only a minor portion of the fan surface. To form and sustain a fan, a fundamental mechanism is essential by which the locus of flow and aggradation is permanently migrating sideways. Avulsion is the essential dominating process of switching active channels to a new path and constructing alluvial fans by repeated coalescing. Basically, alluvial fans should be understood as the product of the merging of diachronous sedimentary piles.

Keywords

Apex · Medial fan zone · Distal zone · Sheetfloods · Abandoned channels · Fan dynamics · Avulsion · Superelevation · Compensational stacking

5.1 Fan Zonation

Fan processes operate on alluvial fans either along channels or in a sheetlike wide manner. The flow regime may vary from flashy episodic to a continuous flow regime. There is a gradual transition from channelized flows at the apex to sheet flows at the toe. The channels are formed either by erosional scour or by levee aggradation.

© Springer Nature B.V. 2019
D. Bowman, *Principles of Alluvial Fan Morphology*,
https://doi.org/10.1007/978-94-024-1558-2_5

From the point of view of the drainage network, alluvial fans can be divided into three adjoining zones (Fig. 1.8):

1. The *apex* or the *proximal zone* is typified by diverging channels. The feeder trunk that enters the fan at the apex expands and branches into numerous parallel, subparallel, and braided channels. Braided channel deposits are marked by horizontal stratification, concave and planar cross-stratification, current ripples, and ripple cross-lamination. The deepest entrenchments on fans occur near the apex and gradually shallow down-fan. Even where fans have merged into a homogeneous *bajada*, the proximal area of channel divergence may still be recognized.
2. The *medial zone* is dominated by bifurcating, parallel, multiple flow paths and overbank sheet flow areas. Reorganization of lateral expanding flows into a parallel braided network is typical. Repeated flow separation occurs by longitudinal and transverse islands and bars with the coarsest gravel along their central axes. Channels branch and separate down-fan more frequently than they join, and the flow is split up into smaller and smaller distributaries of diminishing width, depth, and discharge per channel. Along the down-fan gentler slopes, only small particle sizes can be transported. Confluences of the channels are nearly always graded without breaks of slope. Overprinting of early channel paths by later, stronger flows is common.
3. In the *distal zone*, shallow channels cannot effectively drain flash floods and cause a wide-spreading of overland *sheetfloods* which are thin, relatively uniform flow veneers on the wide floodplain-like distal zone surface, completely unconfined by channel boundaries and typified by overall decreases in depth and gradient. The sheetfloods show a diminishing conveyance capacity, spread layers of gravel and sand, and build a smooth apron of fine sediments on the fan's distal regions. The flood areas, as compared to channeled routes, increase down-fan, indicating that more and more of the total discharge is conveyed by surface sheet flow instead of by the channel network. The three zones of the fan are better distinguishable on larger fans where clearer transitions among the zones are clearer shown.

A typical distributary fan system shows the change from a single-channel at the fan head to a braided-channel flow and sheet flow down-fan. Flow width shows transition from laterally constrained channelized flows with higher transport rates under narrow and deep conditions at the fan apex to wide sheet flows as they stream down-fan. Toward the distal zone, the hydraulic geometry changes (Fig. 3.8), and channels become shallower with a reduced discharge, while channel size, flow depth, and the velocity decline down-fan and channel conveyance drop. Along the fan much water is lost due to infiltration and evapotranspiration. Discharge and stream power may diminish to zero within or slightly beyond the distal zone. During periods of low discharge, the peripheral regions of the fan may become water-starved, even though the proximal zone still receives water.

Apart from zonation, the fan surface can be divided into active and inactive areas. An abandoned, inactive fan surface may evolve: (1) as a result of channel migration away or (2) following deep entrenchment which sets the fan surface above the flood level. Active channels in arid fans show up by their denser vegetation, sharper morphology, and brighter tone. The ratio of active areas to the total fan area indicates the level of the *fan dynamics*.

5.2 The Spread of Activity

Bull (1977) estimated the amount of fan area covered by water during floods. He found that small fans < 1 km^2 could have over 20% of their surface covered by water during floods of low recurrence intervals, whereas fans larger than 100 km^2 had only 5% of their surface covered by water during floods of similar recurrence interval. Active fan areas measured by Millard et al. (2010) ranged from 5% to 8% of the fan area. These findings show that only a small part of the fan area is active at any one time. The smaller the fan and the greater the flood event, the greater will be the fan area contemporaneously activated by a single flood. Most of the fan surface will always be inactive. The widespread channel morphology on fans does not indicate an overall simultaneous fan activity.

An alluvial fan is built by merging deposits named lobes. Fans should be understood as merging of diachronous lobes. The typical alluvial fan shape builds up only over a long stretch of formation time as a cumulative product of multiple depositional events. The fundamental fan-forming process is based on the constant lateral migration of sediment deposition through lobe shifting over the entire fan area (Milana and Tietze 2002). Flow paths must be unstable in order for the entire fan to be deposited and maintained.

5.3 Avulsion

Alluvial fans are the product of a long-term process of shifting channels. *Avulsion* is the term for switching of an active channel to a new path (Ashworth et al. 2004; Field 2001; Mohrig et al. 2000; Jerolmack and Mohrig 2007; Jerolmack and Paola 2007). The new channel path usually follows the steepest descent. *Full avulsion* is one in which all flow is transferred out of the parent channel. Relocation of the active channels can be very local, rejoining the abandoned channel some distance downstream and thus forming an anastomosing pattern. In most cases the channel finds its way back to its previous path somewhere downstream. If the diverted channel does not rejoin the original channel, for example, on megafans, it may abandon its master channel and migrate hundreds of kilometers from its former position. *Nodal avulsion* occurs at the apex and may divert the main part (>50%) of the fan's total discharge (Millard et al. 2010). Small shifts near the fan head can cause dramatic changes in the channel position farther down-fan. Large floods are avulsion-triggering events, making avulsion typically autocyclic and stochastic (Rachocki 1981).

Various conditions and processes have been suggested as those which trigger avulsion (Bryant et al. 1995; Field 2001; Slingerland and Smith 2004; Schumm 2005; Reitz and Jerolmack 2012):

1. Water flows have the tendency to continue their flow direction along the medial line of the fan, thus making it the site of maximum deposition and vertical growth with steeper lateral slopes that invite avulsion by the more favorable downstream gradient advantage. Channels will finally change their path and follow the steeper lateral descent (Fig. 1.7).
2. Channels may avulse due to in-channel aggradation, i.e., due to backfilling following loss of discharge and reduction in capacity. When the channels are oversupplied with sediments and unable to transport their load, deposition drives backfilling from the fan margin toward the apex, and arrested flow may seek new outlets. Backfilling may also occur in the wake of of log jams, vegetative blockages, and debris dams. Deposition of bedload within the channel will accelerate *superelevation* of the channel bed, i.e., the channel becoming perched above its surrounding. During floods of extreme discharge, avulsion will occur from flows that overtop the banks, expanding out of the channel in favor of lateral routes. Rapid aggradation and accumulation of bedload within the channel that accelerates the rise of the channel bed is considered as one of the primary reasons for avulsion.
3. Formation of a mid-channel bar following deposition of a single large boulder or deposition of sieve deposits that initiate clogging and chocking of the channel will divide the flow around the obstruction and intensify lateral bank erosion till channel shift captures the adjacent channel. Channel plugging by debris-flow deposition may similarly cause the flow to divert. Tectonic tilting may also contribute toward a sideway shift.
4. *Headward gully erosion* moving upward the fan surface, triggered by a lowering base-level, may contribute toward avulsion by channel piracy.

Avulsion causes many channels on the fan surface to become *blind* or *abandoned* (Field 2001), i.e., cut off and become disconnected from the active drainage system, draining only local rainwash. During great flooding events, abandoned channels provide ready routes for the diverted flows (Mohrig et al. 2000; Jerolmack and Mohrig 2007). Such *reoccupation* of pre-existing abandoned channels on the fan surface is frequent (Beaty 1963). Reoccupation of channels indicates that the inherited topography acted as an attractor to avulsing channels. Reoccupied channels often reincise, become wider, and develop vertical banks. Channels that changed their paths imprinted in the stratigraphic record an architecture of stacked sand or gravel bodies (Jerolmack and Paola 2007). Unconformities within the fan stratigraphic sequence are often also the outcome of lateral switching of channels (Harvey 1987).

The frequency of channels' switch into new flow pathways is defined as *avulsion frequency*, which increases with increase in the ratio of sediment supply to water discharge. The timescale of channel avulsion can be estimated in terms of the volume to be filled with sediment divided by the rate at which sediment enters the system (Reitz and Jerolmack 2012). Aggradation and backfilling promote increased channel mobility, bifurcation, and avulsion frequency, which increases faster than linearly with sediment feed rate (Bryant et al. 1995). The avulsion cycle operates on timescales of centuries to millennia and also constructs large fans such as those of the Kosi. The avulsion process may be allogenically controlled in case its frequency shows the periodicity of an allogenic factor.

Shifting of channels by lateral migration is a must in order to form alluvial fans by coalescing. The paleosols that develop during channel shifts tend to be less mature and less laterally extensive because of shorter exposure periods compared to soils formed on an abandoned fan surface following deep incision. Protection programs to restrict channel migration include building channel-confining levees and straightening, deepening, and widening channels in order to increase their conveying capacity. Due to the lateral depositional and incisional activity, no locality on a fan preserves the full continuous evolutional record, and nowhere on an alluvial fan can a vertical stratigraphic succession exhibit the entire fan history. On hyperarid alluvial fans, where the distribution of vegetation closely follows the channels, relocation and diversion of channels are major reasons for mortality of trees that have no deep root system, such as *Acacia*. The spatial pattern of the tree mortality often reflects the dependence of trees on the stability of the surface water routes.

5.4 Compensational Stacking

Flow and deposition are influenced by the locations of previous elements. The *compensational stacking* tendency (Straub and Pyles 2012), i.e., when flow deposits fill topographic lows and smooth out topographic relief in sedimentary basins, was related to turbidities and to the fluvial architecture in deltas (Stow and Johansson 2000; van Wagoner 1995; Straub et al.2009). The compensational stacking depositional trend is interpreted to minimize the potential energy associated with elevation gradients (Stow and Johansson 2000; Straub et al. 2009; Straub and Pyles 2012). During this process, relict abandoned channels, when becoming lows, are expected to serve as attractors (Jerolmack and Paola 2007) and become gradually filled. The local deposit thickness is balanced, at least partly, by the compensational behavior of the flow. The compensational tendencies increase in strength with larger-scale events.

The compensational stacking concept seems relevant to fans, although it was not mentioned when the architecture of the fan stratigraphy had been described. According to the compensational stacking concept, an alluvial fan is a large depositional unit, composed of smaller lobe elements deposited variably temporally and spatially. The lobes and the fans are scale variants with boundaries that control avulsion which redirect the sediment-transport to local topographic lows.

References

Ashworth PJ, Best JL, Jones M (2004) Relationship between sediment supply and avulsion frequency in braided rivers. Geology 32:21–24

Beaty CB (1963) Origin of alluvial fans, White Mountains, California and Nevada. Ann Assoc Am Geogr 53:516–535
Bryant M, Falk P, Paola C (1995) Experimental study of avulsion frequency and rate of deposition. Geology 23:365–368
Bull WB (1977) The alluvial-fan environment. Prog Phys Geogr 1:222–270
Field J (2001) Channel avulsion on alluvial fans in southern Arizona. Geomorphology 37:93–104
Harvey AM (1987) Alluvial fan dissection: relationships between morphology and sedimentation. In: Frostick LE, Reid I (eds) Desert sediments: ancient and modern. Geological Society Special Publication, vol 35. Blackwell, Oxford, pp 87–103
Jerolmack DJ, Mohrig D (2007) Conditions for branching in depositional rivers. Geology 35:463–466
Jerolmack DJ, Paola C (2007) Complexity in a cellular model of river avulsion. Geomorphology 91:259–270
Milana JP, Tietze KW (2002) Three-dimensional analogue modelling of an alluvial basin margin affected by hydrological cycles: processes and resulting depositional sequences. Basin Res 14:237–264
Millard TH, Hogan DL, Wilford DJ, Roberts B (2010) A method to assess fluvial fan channel networks, with a preliminary application to fans in coastal British Columbia. Geomorphology 115:286–293
Mohrig D, Heller PL, Paola C, Lyons J (2000) Interpreting avulsion process from ancient alluvial sequences: Guadalupe-Matarranya system (northern Spain) and Wasatch Formation (western Colorado). Geol Soc Am Bull 112:1787–1803
Rachocki A (1981) Alluvial fans. Wiley Interscience, New York, 161 p
Reitz MD, Jerolmack DJ (2012) Experimental alluvial fan evolution: channel dynamics, slope controls and shoreline growth. J Geophys Res Earth Surf 117(F2):F02021. https://doi.org/10.1029/2011JF002261
Schumm SA (2005) River variability and complexity. Cambridge University Press, Cambridge, UK
Slingerland R, Smith ND (2004) River avulsion and their deposits. Rev Earth Planet Sci 32:257–285
Stow DAV, Johansson M (2000) Deep-water massive sands: nature, origin and hydrocarbon implications. Mar Pet Geol 17:145–174
Straub KM, Pyles DR (2012) Quantifying the hierarchical organization of compensation in submarine fans using surface statistics. J Sediment Res 82(11):889–898
Straub KM, Paola C, Mohrig D, Wolinsky MA, George T (2009) Compensational stacking of channelized sedimentary deposits. J Sediment Res 79(9):673–688
Van Wagoner JC (1995) Sequence stratigraphy and marine to non-marine facies architecture of foreland basin strata, Book Cliffs, Utah, U.S.A. In: van Wagoner JC and Bertram GT (eds) Sequence stratigraphy of foreland basin deposits: outcrop and subsurface examples from the cretaceous of North America: American Association of Petroleum Geologists, pp 137–223

6 Flows

Abstract
Flows on alluvial fans fall within the continuum from Newtonian water flows to debris flows as end members. Hyperconcentrated flows are the transitional stage. Newtonian water flows behave as turbulent nonhomogeneous flows with the fluid, and the granular phases free to act independently. Intermediate hyperconcentrated flows are characterized by very high sediment concentrations which damp turbulence and decrease the settling velocity. Debris flows are slurries composed of disorganized cobbles and boulders embedded in a matrix as a single-phase mixture. Debris flows may be more dilute or more viscous, depending on the proportions of the fines and the granular components. The field characteristics of the debris flows are, among others, the matrix-supported framework, poor sorting and lack of stratification or other sedimentary structures, and a lobate snout at the flow front and lateral levees. Due to its mechanical strength, the viscous mass of debris flows supports the particles and provides great transporting competence, up to huge boulders weighting several tons.

Keywords
Newtonian water flows · Hyperconcentrated flows · Debris flows · Transporting competence · Mobility · Levee

6.1 Water and Hyperconcentrated Flows

Flows on alluvial fans fall within the continuum from fluvial water flows to debris flows as end members, including hyperconcentrated floods as a transitional stage described by Costa (1988). *Newtonian water flows* behave as turbulent flows without appreciable shear strength, have a relative low sediment concentration of up to 40% by weight, and have bulk densities in the range of 1.01–1.33 g/cm^3 (Pierson and Costa 1987). In Newtonian mixtures any sand and gravel particles present are

© Springer Nature B.V. 2019
D. Bowman, *Principles of Alluvial Fan Morphology*,
https://doi.org/10.1007/978-94-024-1558-2_6

free to settle out of suspension. The mixtures are nonhomogeneous because the fluid and granular phases are free to act independently. Water flows frequently show field velocities in the range of 4–8 m/s. The larger drainage basins export flows of higher water/sediment ratios, i.e., are more fluvially dominated. Intermediate *hyperconcentrated flows* are characterized by very high sediment concentrations, 40–70% solids by weight (Costa 1984; Pierson 2005), which damp turbulence and decrease settling velocity. Sediment/water mixtures maintain Newtonian behavior while retaining the water and the sediment phases separated and free to act independently. That the mixing of sediment in the water is not complete is demonstrated by the rapid increase in solids concentration toward the bed. The clay and much of the silt are, however, considered to be an intrinsic part of the fluid because they usually will not settle out of suspension during a natural flow event. Hyperconcentrated flow deposits are thinner than debris flows and may show alternating clast and matrix-supported frameworks, moderate sorting, and initial stratification. The mass density of hyperconcentrated flows is in the 1.3–1.8 g/cm^3 range, transitional between water and debris flows.

6.2 Debris Flows

Characteristics Alluvial fans are usually sourced by coarse sediments. Mudflows which are composed entirely of silt and clay are not a typical fan component. Debris flows (Costa 1984) are non-Newtonian slurries composed of masses of disorganized cobbles and boulders embedded in matrix as a single-phase mixture, making a quasi-plastic mass. When the shear strength of the mixture becomes sufficient to suspend gravel size, the particles are not able to settle independently, and the flowing mixture becomes a debris flow typified by lack of stratification or other sedimentary structures, matrix-supported clasts, and very poor sorting. The largest flows can transport outsized boulders, 10 m or more in diameter (Pierson and Scott 1985).

The typical debris flow behaves as a single-phase mixture wherein the solids constitute 70–90% by weight, mass densities are in the range of 1.8–2.5 g/cm^3, and the amount of water may range from 15% to 60% by weight. The high density of debris flows reduces settling velocity, and thus coarser material can be maintained longer in suspension. The clasts are dispersed and chaotically arranged, lacking any degree of grading and providing internal friction that contributes to the shear strength. Debris flows may be more dilute or more viscous, depending on the proportion between fines and granular components. When the mass becomes saturated, pore pressure increases and shear strength decreases, transforming the material from rigid into a plastic-viscous mass (Johnson 1970).

Debris flows are matrix-supported with mud surrounding the large particles. Low abundance of fines and in particular clay prevents transformation of sediments into debris flows. Debris flows range from clast-rich to mud-rich. Matrices are typically composed of an unsorted mixture of granules, pebbles, and sands in different proportions and with 5–15% silt and clay by weight (Lowe 1979; Nemec and Steel

1984; Whipple and Dunne 1992). Flows with the highest sediment/water ratios and, therefore, of great strength come to rest on relatively steep gradients. Where debris flows have more substantial amounts of silt or clay, they may be mobile even on very low slopes (e.g., <0.01; Stock 2013). The most dilute debris flows do travel as far as the lower parts of the fans. From the steepness point of view, alluvial fans are not the optimal steep environment to expect debris-flow activity, except at the apex.

Sourcing and Controlling Conditions Fans that are composed dominantly of debris flows are associated with small and steep source basins. Smaller and steeper basins tend to produce more debris flows because of greater erosion and sediment supply per unit runoff-producing area (Fig. 6.1), whereas less steep, larger basins

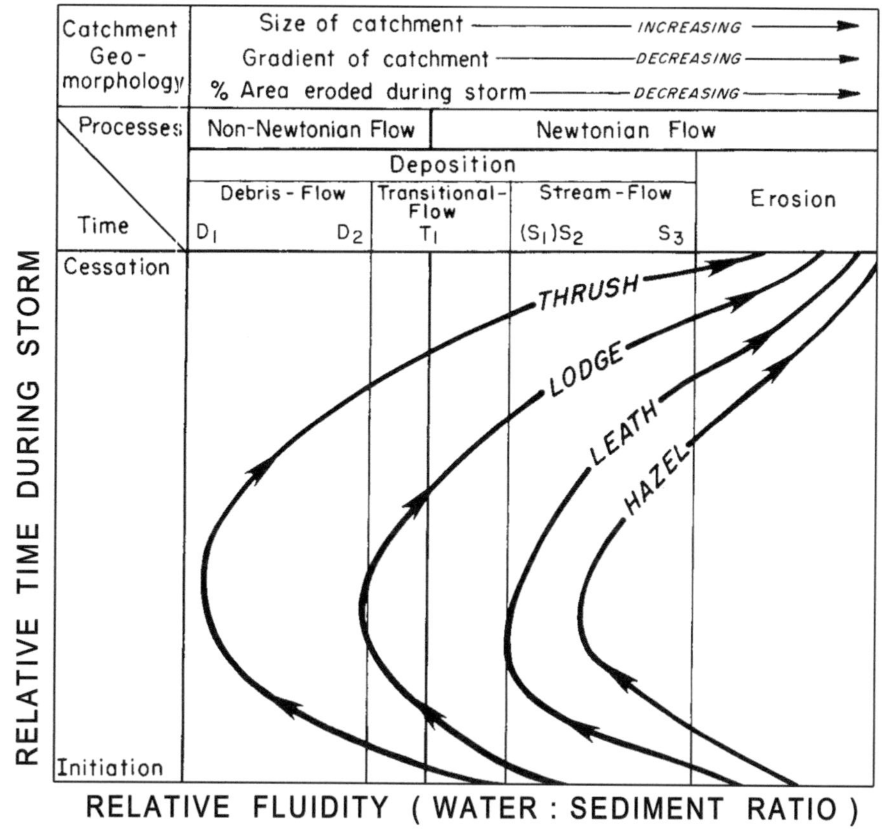

Fig. 6.1 Conceptual model showing the changes in the water/sediment ratio, the sequence of depositional and erosional events and the associated flow conditions during the 1982 storm for four study fans in the Howgill Fells. Shown are the relative timing of facies deposition and the amount of erosion: fan with a smaller catchment, steeper catchment channels and larger percentage of area eroded in the basin (Thrush) experiences more time of debris and transitional flows relative to a fan (Hazel) with larger catchments, shallower channel slopes and less percentage of eroded area. (Redrawn from Wells and Harvey 1987)

tend to produce streamflow deposition (Kostaschuk et al. 1986; Wells and Harvey 1987). The debris flows generated on the steeper upper slopes in a large basin may terminate in the gentler reaches before spreading down to the fan. In order to keep moving, debris flows require gradients that are steeper than those of water flows, often 12–17° and up to 25° steep, typical to slopes of the first- and second-order basins which get relatively more rain, and are the most prone to landslides.

Melton's (1965) ruggedness number $H_d/A_d^{0.5}$, where H and A are the basin height and area above the fan, respectively, is successful in differentiating between basins supplying debris flows and water floods. In small and steep basins of high Melton number, debris flows gain a longer runout and are able to reach the fan. Debris flows can also be deposited directly from the mountain front on the fan. A low vegetation cover or destruction of vegetation may contribute toward triggering debris flows. After forest fires, devegetation and decay of roots anchoring colluvium increase sediment availability and promote debris flows. Regrowth of plant cover after a fire will reduce the slope sediment yield (Meyer and Wells 1997).

Superposition of debris-flow beds allows estimation of their recurrence interval by dating. Recurrence intervals of about 300 years have been reported (Beaty 1974; Hubert and Filipov 1989) as well as intervals of 3000–4000 years (Eaton et al. 2003). A sufficiently long relatively dry interval is needed to prepare the deeply weathered rock unit for next transportation.

The first condition for the formation of debris flows is the *availability of weathered coarse material*. Accumulation and buildup of debris in the sourcing basin must exceed a critical thickness needed to trigger debris flows (Heward 1978). Additionally, debris-flow initiation needs water (Costa 1984). A critical rainfall threshold must be reached based on rainfall intensity and a minimal duration. Rainfall events with intensities of a few tenths of mm/h and a total of a few hundreds of mm may trigger shallow *landslides* that are a potential source of debris (Zanchettaa et al. 2004). Weathering-limited basins exhibit a lower frequency of debris-flow activity. A sediment source which provides abundant muddy material for the matrix, as pyroclastics or eolian silty-clayey deposits, provides favorable conditions. Too little fine material in the source area inhibits debris-flow formation. Less than 10% of clay and silt may be enough for debris flows to form, given a long enough recharge period. After events of large inputs of debris flows, sediment becomes less available, and the conditions become more dilute. The transported clast size decreases, and stratification and the clast support improve. With time, later floods remobilize the earlier deposited finer-grained sediments, leaving the surface of debris flows as an armor of coarse clasts (Wells and Harvey 1987).

Mobility Silt and clay control the retention of water and thus the mobility of debris flows. The resistance to motion results from the viscosity and the high shear strength which must be exceeded before any flow can be initiated. Failure takes place when strength τ of the water-solid mix is exceeded. The increase of pore-fluid pressure reduces the shear strength, and the increase of fines enables debris flows to move along gentler slopes. The lower the viscosity, the far off debris flows may prograde.

Excessive infiltration following long intensive rain with an intensity of about 60–90 mm/h is an important trigger to mobilize material. Not too frequent precipitation events will leave a too long dry period. Under contrasting conditions, too large volumes of water, relative to the unconsolidated debris, will produce water flows.

To be mobile, debris flows with low water content need steeper fan slopes (4–8°), typically found in the proximal parts. Debris flows proceed to the midfan and usually do not reach the toe. The upper fan zone exhibits deposition of more viscous debris flows of higher internal shear strength that halt at higher gradients. Distal fan locations display thinner debris flows with lower internal shear strength (Staley et al. 2006). Debris flows move at speeds of <1–20 m/s (Leopold et al. 1964; Takahashi 1980). Loss of water following fluid escape and increase in the internal pore friction to a point that the shear stress becomes exceeded by shear strength are the main reasons for debris flow to stop. Shallower sheets increase the loss of water and thus the internal pore friction.

Boulders in debris flows are supported by the fine matrix. The cohesive strength of clay-water slurries with densities of 1.17–1.26 g cm^3 is generally proportional to the clay content (Pierson 1981). High density reduces the settling velocity, and coarse material can be maintained longer in suspension. The viscous mass, due to its mechanical strength, supports the particles and provides the great *transporting competence*, up to huge boulders of tons weight.

The thicker the debris flow, the larger the particle it can carry. Thickness thus correlates strongly with maximum particle size and is a measure of competence. The larger particles tend to drift toward outside edges, i.e., to move to the top and front of flows. Deposition of the coarsest fragments (Bull 1964) will increase the relative water content and initiate the transformation to a more dilute hyperconcentrated flow (Pierson and Scott 1985). The debris flows are not only effective in entraining loose channel sediment but also at eroding the bedrock. Impact-related wear during the passage of large entrained particles can play a primary role in eroding bedrock in steep channels. Thin bed cover negates the effectiveness of particle impacts in damaging subjacent bedrock (Stock and Dietrich 2003, 2006). The debris flows will stop moving, i.e., start to *freeze* and come to rest at their distal parts and continue freezing backward.

Morphology Debris flows typically extend from the fan head down to the middle-fan zone but may also spread along the entire radial extent of the fan. The debris flows may settle in narrow strips, several hundred meters long and a few tens of m wide, but may also form several km long sheets. Debris flows become shallower with spreading, typically running out from about 5 m thickness to a few decimeters. The *runout* length of a debris flow is basically controlled by its volume and viscosity, whether muddy or more granular, as well as by the slope and topography of the fan surface. Debris flows tend to smoothen the surface of the fan by spreading into thin sheets, filling channels and surface undulations, and burying the preexisting topography, thus interfering with avulsion and channel shifting (Whipple and Dunne 1992). Debris flows may, however, also ignore the former fluvial channel system.

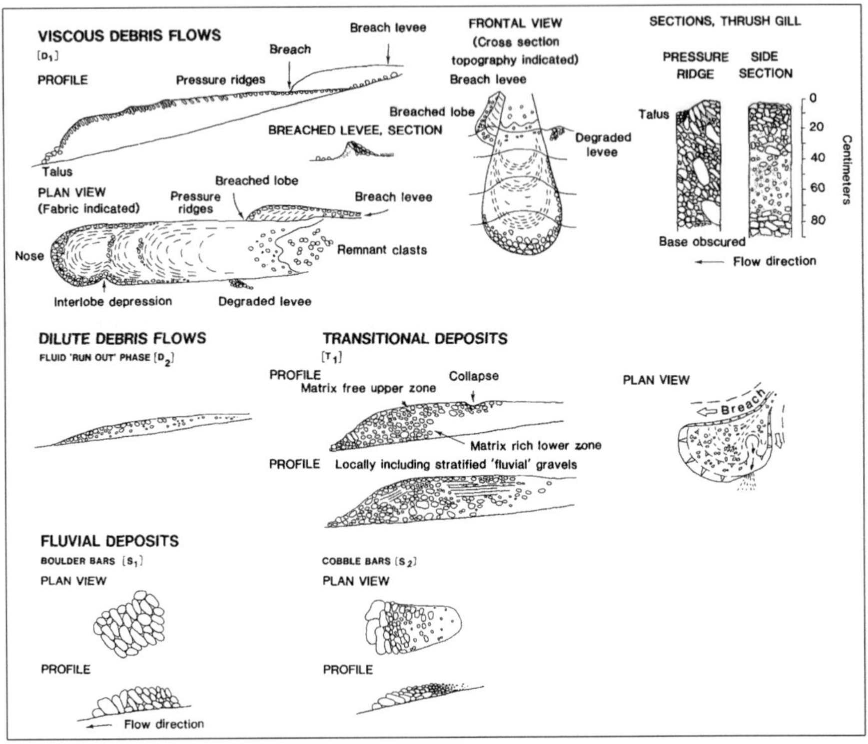

Fig. 6.2 Geomorphic and sedimentologic features of major debris flow facies types: **(D1)** viscous debris flows-pressure ridges, levees, stratigraphy and lobes; **(D2)** dilute debris flows; **(T1)** transitional flows**;** and **(S1, S2)** fluvial bars. (Following Wells and Harvey 1987)

Debris flows have typically lobate, tongue-shaped form with sharp margins (Fig. 6.2). The steep frontal snout, which consists of coarse clasts, may look as a boulder dam with steep flow margins, pushed from behind by a more fluid mud slurry (Savage and Iverson 2003). Piled-up coarse material forms *boulder lobes* which often lend to the fan surface a typical *hummocky* appearance. The moving debris flows push aside the slower-moving flanks and form a pair of parallel, elongated ridges, called *levees*, up to a few m tall and a few tenths of m wide. The space between levees consists of a typical trapezoidal-shaped channel. The lobe tops show a *pressure ridge morphology* with concentric bands of elongated clast, oriented parallel to the lobe front (Fig. 6.2). The coarsest particles in the mixture are displaced and segregated toward the surface and the snout and toward the lateral edges of the flow (Hubert and Filipov 1989).

On forested fans (Wilford et al. 2005), scars on tree stems may indicate abrasion by transported boulders or logs and provide an indication of the thickness of debris flows. Large boulders may be wedged between trees, 1–2 m or more above the ground surface. Deposition of sediments behind logs forms a microscale stepped profile. Multiple tree stems lodging against standing trees may trap sediments, over

4 m in height on their upslope side. Accumulation of small woody debris against streamside vegetation helps to confine and keep the water in the channel, maintain the stream power, and limit channel avulsions. In forested fans, drainage routes can, however, become obstructed with sediment and debris, leading to channel avulsion. Downed logs, tree stems, and shrubs may act to disperse flow, reduce velocities, and increase the hydraulic roughness.

References

Beaty CB (1974) Debris flows, alluvial fans, and a revitalized catastrophism. Z Geomorphol 21:39–51

Bull WB (1964) Geomorphology of segmented alluvial fans in western Fresno County, California. US Prof Pap 352-E:89–129

Costa JE (1984) Physical geomorphology of debris flows. In: Costa JE, Fleisher PJ (eds) Developments and applications of geomorphology. Springer, Berlin, pp 268–317

Costa JE (1988) Rheologic, geomorphic and sedimentologic differentiation of water floods, hyperconcentrated flows and debris flows. In: Baker VR, Kochel RC, Patton PC (eds) Flood geomorphology. Wiley, New York, pp 113–122

Eaton LS, Morgan BA, Kochel RC, Howard AD (2003) Quaternary deposits and landscape evolution of the central Blue Ridge of Virginia. Geomorphology 56:139–154

Heward AP (1978) Alluvial fan and lacustrine sediments from the Stephanian A and B (La Magdalena, Cinera-Matallana and Sabero) coalfields, northern Spain. Sedimentology 25:451–488

Hubert JF, Filipov AJ (1989) Debris-flow deposits in alluvial fans on the west flank of the White Mountains, Owens Valley, California, U. SA Sediment Geol 61:177–205

Johnson AM (1970) Physical processes in geology. Freeman, Cooper, San Francisco, 577 pp

Kostaschuk RA, MacDonald GM, Putnam PE (1986) Depositional process and alluvial fan-drainage basin morphometric relationships near Banff, Alberta, Canada. Earth Surf Process Landf 11(5):471–484

Leopold LB, Wolman MG, Miller JP (1964) Fluvial processes in geomorphology. Freeman, San Francisco

Lowe DR (1979) Sediment gravity flows: their classification and some problems of its application to natural flows and deposits. In: Doyle LJ, Pilkey OH Jr (eds) Geology of continental slopes. Spec Publs Soc Econ Paleont Miner 27: 75–82

Melton MA (1965) The geomorphic and paleoclimatic significance of alluvial deposits in southern Arizona. J Geol 73:1–38

Meyer GA, Wells SG (1997) Fire-related sedimentation events on alluvial fans, Yellowstone National Park, U.S.A. J Sed Res 67:776–791

Nemec W, Steel RJ (1984) Alluvial and coastal conglomerates: their significant features and some comments on gravelly mass-flow deposits. In: Koster EH, Steel RJ (eds) Sedimentology of gravels and conglomerates. Can Soc Pet Geol Mem 10:7–31

Pierson TC (1981) Dominant particle support mechanisms in debris flows at Mt Thomas, New Zealand, and implications for flow mobility. Sedimentology 28:49–60

Pierson TC (2005) Hyperconcentrated flow – transitional process between water flow and debris flow. In: Jacob M, Hung O (eds) Debris-flow hazards and related phenomena. Springer, Berlin, pp 159–202

Pierson TC, Costa JE (1987) A rheologic classification of subaerial sediment-water flows. Rev Eng Geol 7:1–12

Pierson TC, Scott KM (1985) Downstream dilution of a lahar: transition from debris flow to hyperconcentrated stream flow. Wat Res Res 21:1511–1524

Savage SB, Iverson RM (2003) Surge dynamics coupled to pore-pressure evolution in debris flows. In: Rickenmann D, Chen C (eds) Debris-flow hazards mitigation: mechanics, prediction, and assessment. Millpress, Rotterdam, pp 503–514

Staley DM, Wasklewicz TA, Blaszczynski JS (2006) Surficial patterns of debris flow deposition on alluvial fans in Death Valley, CA using airborne laser swath mapping data. Geomorphology 74(1):152–163

Stock JD (2013) Waters divided: a history of alluvial fan research and a view of its future. In: Wohl E (ed) Treatise on geomorphology, Fluvial Geomorphology, vol 9. Academic, San Diego, pp 413–458

Stock JD, Dietrich WE (2003) Valley incision by debris flows: evidence of a topographic signature. Water Resour Res 39:1089

Stock JD, Dietrich WE (2006) Erosion of steepland valleys by debris flows. Geol Soc Am Bull 118(9–10):1125–1148

Takahashi T (1980) Debris flow on a prismatic open channel. ASCE J Hydraul Div 106:381–396

Wells SG, Harvey AM (1987) Sedimentologic and geomorphic variations in storm-generated alluvial fans, Howgill Fells, northwest England. Bull Geol Soc Am 98:182–198

Whipple KX, Dunne T (1992) The influence of debris-flow rheology on fan morphology, Owens Valley, California. Geol Soc Am Bull 104:887–900

Wilford DJ, Sakals ME, Innes JL, Sidle RC (2005) Fans with forests: contemporary hydrogeomorphic processes on fans with forests in west central British Columbia, Canada. Spec Publ Geol Soc Lond 251:25–40

Zanchettaa G, Sulpizioa R, Di Vito MA (2004) The role of volcanic activity and climate in alluvial fan growth at volcanic areas: an example from southern Campania (Italy). Sediment Geol 168:249–280

7 Aggradation

Abstract

The overall fan evolution pattern consists of a primary depositional stage followed by a secondary stage of dissection. Net deposition is conditioned by a high uplift rate of the mountainous block, leaving the feeding channel at the mountain exit above the uppermost fan level. Aggradation is controlled mainly by the change in the hydraulic geometry, by water loss due to percolation and following an increase in the sediment size and amount. Fans are products of merging lobes, making the fan a diachronous unit. The marginal and distal fan sediments interfinger with those of the neighboring environments while thriving toward a mutual adjustment.

Keywords

Hydraulic geometry · Overbank deposition · Sheetflood deposits · Sieve deposits · Lobe · Backfilling · Interfingering

Blair and McPherson (2009) distinguish between a *primary* and a *secondary* stage in fan development. The primary processes are aggradational, i.e., those that deposit the sediments supplied from the drainage basin. Later, secondary processes dissect or destroy the fan. The basic conditions for fan deposition are (1) high uplift rate of the source province and (2) relative hard source rocks, allowing a relatively slow rate of the drainage basin entrenchment, preserving the sourcing channel at the mountain exit above the fan (Bull 2007).

Under conditions of low-discharge events with high rates of water loss in the drainage basin and insufficient gradients of the feeder, aggradation will start to build up a steep fan close to the mountain front, while the down-fan zone may be starved of water and of sediment supply. Experimental data indicate that high discharge enhances proximal erosion and accumulation at mid and distal sites. During low discharge, deposition moves upstream as backfilling. Quantitative estimations of

© Springer Nature B.V. 2019
D. Bowman, *Principles of Alluvial Fan Morphology*,
https://doi.org/10.1007/978-94-024-1558-2_7

aggradation can be obtained based on the difference in elevation over a period, calculated from digital terrain models (DTM).

Fans grow in volume when more sediments arrive at the fanhead than are removed from the fan's toe (Milana and Tietze 2002). The rates of increase in the fan area and in its thickness become smaller as the fan area grows. Fan volume is stabile when there is no net aggradation, i.e., when the rate of sediment supply equals the rate of removal. Net aggradation occurs when none of the active channels reach the down-fan axial basin nor any down-fan water body, and all sediments are deposited on the fan. In such evolutionary phase, the alluvial fans are disconnected and uncontrolled by the base-level. When sediments bypass the fan, i.e., material is transported through the fan system and beyond, the down-fan base-level will start to control the fan.

Aggradation on alluvial fans is controlled mainly by the change in the hydraulic geometry, by water loss due to percolation and following increase in the sediment size and amount. The following parameters determine the rate of aggradation:

(a) *Change in the hydraulic geometry* – The creek that emerges from the confinement of the mountains changes at the mountain outlet from channelized to a widely, radially dispersed, shallow, and braided drainage network, triggering thereby a dramatic increase in flow width and a decrease in depth, thus causing the wetted perimeter, i.e., the interface between the liquid and the boundary which confines the fluid, to increase. The hydraulic radius and Shields' stresses decrease down-fan (Fig. 3.8; Stock et al. 2007). The result is an increasing hydraulic roughness, causing a fall in the velocity and an immediate drop in stream power and in the flow capacity, which more than offsets the sudden increase in gradients of upper fan channels compared to the upstream trunk. At the lower down-fan area with its gentler slopes, the entrainment stresses progressively decline beyond threshold. The wetting of the down-fan zone also increases the density of the vegetation, which, by resisting the flow, increases the hydraulic roughness (Bull 1997).

Following the incapability of channels during major floods to contain the discharge, channels often fill up and overflow, causing extensive *overbank deposition* in the middle and lower fan zones. These laterally wide and unconfined sheet flows are shallow, a few decimeters deep and supercritical (Blair and McPherson 1994; Parker et al. 1998). Their deposits which share a relative well-sorted and clast-supported character (Wells and Harvey1987; Costa 1988) consist of mud, sand, some granules, and fine pebbles, with the grain size decreasing downslope and away from the distributaries. These *sheetflood deposits* are planar, composed of rhythmic couplets of alternating sand and clast-supported gravel, oriented subparallel to the fan surface with well-developed imbrication. Sheetflood deposits dominate the lower and distal parts of alluvial fans, with silt becoming dominant downslope. Catastrophic unconfined sheet flooding of exceptional magnitude may also deposit coarse gravel to boulders on the abandoned, inter-channel fan surfaces that are a few meters above the active entrenched ephemeral drainage system (Wells and Dohrenwend 1985).

(b) *Water loss due to percolation* – High rates of water percolation into the permeable fan gravel cause loss of discharge and initiate sediment deposition. When much discharge is lost, the outer parts of the fan system receive little or any water. The permeable fan surface acts as a sieve into which the water infiltrates, leaving a surface residue of coarse, non-sorted, clast-supported, open-framework, matrix-poor deposits which may include outsized cobbles and boulders (Hooke 1967). These *sieve deposits* form a characteristically hummocky surface. Sieve deposits may also form following winnowing and removal of fine matrix out of debris flows. The fan surface deposits, including sieve deposits, form tongue-like *lobes* radiating down-fan from individual delivery channels (Schumm et al. 1987). Lobes are smaller fan-shaped deposits that prograde from the outlet of active incised channel, typically at intersection points, and are the basic deposits composing alluvial fans. Alluvial fans are the product of merging lobes of different ages.
(c) *Significant increase in the sediment supply*, without a concomitant increase in the overall stream power, will cause *backfilling*, i.e., in-channel aggradation which reduces the height of the channel banks through a rise in the channel bed elevation. Such *channel refilling* may raise the channel bottom by several meters and may be followed later by re-entrenchment into the deposited fill.

Neighboring Environments The contiguous environments of alluvial fans are prograding talus sliding from the mountain front, mainly by the fan head; laterally adjacent fans; interfan areas; a river or floodplain along the axial basin; a playa or a dune field at the distal fan margins; and a lake or a seashore. A trunk river at the base of a fan may trim the toe. Otherwise, the fan may block the trunk river.

The contact between the environments is dynamic, thriving toward a mutual adjustment while controlled by the relative rates of deposition and usually shown as *interfingering*, i.e., interbedding with the neighboring environments. The interactions will result in mixed facies along the interface, shown in the stratigraphy as well-bedded lacustrine sediments within gravelly fan deposits, or coastal swash bars on top or between alluvial fan sediments, or by fan deposits interbedded with aeolian sands.

When an alluvial fan progrades toward a former lake bottom, it will bury the lake sediments, and with a returning rise of the lake, the distal fan will be flooded and buried by the lake margin facies, although the proximal zone will remain unaffected. The subsequent last withdrawal of the lake will be marked by wave-cut shorelines and gravel beaches, characterized by subrounded to rounded, well or moderately sorted gravel. Later, fan activity will truncate or burry the raised shorelines.

References

Blair TC, McPherson JG (1994) Alluvial fans and their natural distinction from rivers based on morphology, hydraulic processes, sedimentary processes and facies assemblages. J Sediment Res 64:450–489

Blair TC, McPherson JG (2009) Processes and forms of alluvial fans. In: Parsons AJ, Abrahams AD (eds) Geomorphology of desert environments, 2nd edn. Springer, New York, pp 413–467

Bull WB (1997) Discontinuous ephemeral streams. Geomorphology 19:227–276

Bull WB (2007) Tectonic geomorphology of mountains: a new approach to paleosiesmology. Blackwell, Oxford

Costa JE (1988) Rheologic, geomorphic and sedimentologic differentiation of water floods, hyperconcentrated flows and debris flows. In: Baker VR, Kochel RC, Patton PC (eds) Flood geomorphology. Wiley, New York, pp 113–122

Hooke RL (1967) Processes in arid-region alluvial fans. J Geol 75:438–460

Milana JP, Tietze KW (2002) Three-dimensional analogue modeling of an alluvial basin margin affected by hydrological cycles: processes and resulting depositional sequences. Basin Res 14:237–264

Parker G, Paola C, Whipple KX, Mohring D (1998) Alluvial fans formed by channelized fluvial and sheet flow. I: theory. J Hydraul Eng 124:985–9959

Schumm SA, Mosley MP, Weaver WE (1987) Experimental fluvial morphology. Wiley, New York, 413 P

Stock JD, Schmidt KM, Miller DM (2007) Controls on alluvial fan long-profiles. Geol Soc Am Bull 120:619–640

Wells SG, Dohrenwend JC (1985) Relict sheetflood bed forms on late quaternary alluvial-fan surfaces in the southwestern United States. Geology 13:512–516

Wells SG, Harvey AM (1987) Sedimentologic and geomorphic variations in storm-generated alluvial fans, Howgill fells, Northwest England. Bull Geol Soc Am 98:182–198

8 Fan Entrenchment

Abstract

When erosional processes start to control and reshape the fan, its original aggradational form is gradually modified. The power to erode is controlled by the absolute mass density, by the mean flow depth, by the gravity gradient, and by the flow velocity. All variables, if increased, favor the transportation of sediments. Channelized flows attain greater flow depth and higher shear stress enabling more efficient sediment transport.

Water/sediment discharge ratio determines whether stream power exceeds the critical power required to enable erosion. High water/sediment ratio that enables erosion may result from entrapping sediments or when fan sourcing comes to an end as a product of aging. For any streamflow and sediment load, there is a threshold slope at which entrenchment will be initiated.

Fan surface entrenchment ranges from shallow fan-head trenching which is prone to backfilling, to deep, through-fan entrenchment which leaves the fan surface inactive. The entrenched channels are more moderate-sloped, reflecting the trend toward reestablishment of grade. Lateral erosion and widening of deeply incised channels may eliminate with time large sections of the fan.

Keywords

Threshold of critical power · Sediment/water discharge ratio · Threshold slope · Fan aging · Shallow fan-head trenching · Through-fan entrenchment · Lobe · Coupling · Intersection point · Telescopic fan

8.1 The Switch to Entrenchment

When the main stage of fan deposition ends, erosion becomes dominant by detachment and transport processes, and the original depositional fan morphology starts gradually to wear down. At this stage the fan surface no longer reflects the

© Springer Nature B.V. 2019
D. Bowman, *Principles of Alluvial Fan Morphology*,
https://doi.org/10.1007/978-94-024-1558-2_8

depositional history but mainly the erosive phase. In the erosional phase, the former depositional, formative processes should be inferred mainly from the stratigraphy.

By directing the flow into confined narrow channels which afford increased *stream power* (Bull 1979, 1988), dispersion and deposition are contra-acted. The stream power is primarily dependent on the *channel geometry*, on the mean boundary shear stress, on the absolute mass density per volume, on the mean flow depth, and on the slope gradient. All variables, if increased, favor the transportation of sediments. Vegetation increases the hydraulic roughness, thus dissipating flow energy and counteracting entrenchment.

Channelized flow is characterized by increased flow depth and higher mean shear stress which enhance sediment transport rates and allow channels to promote a longer travel distance and sediments to be transferred more efficiently down-fan. For the same discharge, more sediment can be carried in channelized, deep, and narrow conditions, relative to wide shallow sheet flows (Parker et al. 1998). Being more efficient, channelized conditions can move sediments over lower slopes and impose lower gradients on fans. Coarse sediments on fans often reflect the high-energy conditions of incised channels that transfer sediments further basinward.

The magnitude of the flood event is a major control of the stream power. Flow velocities on fans may range as high as 9 m/s. The capacity of streams to transport sediment increases with discharge as a power function with an exponent between 2.0 and 3.0 (Leopold et al. 1964; Knighton 1984). Bedload transport rates may, however, be lowered due to extensive armoring of the streambed (Laronne and Reid 1993; Jones and Frostick 2008). This would explain less incision than expected during a period of increased discharge.

Critical power is the stream power required to move sediments. The *threshold of critical power* separates the modes of deposition and erosion in fluvial systems. When stream power/critical power exceeds 1, flood events trigger entrenchment. Stream power is dissipated by the fluid's viscosity, by turbulence, by friction against the channel bottom and the banks, and by the work invested in moving the bedload and maintaining the flow against resistance. In narrow and deep channels, where the stress for sediment movement exceeds the threshold of detachment, the coarser sediments will be transported, and the channel will be deepened by scouring.

8.2 Controls of Entrenchment

Sediment/Water Discharge Ratio Fluctuating sediment supply, combined with cyclical hydrological fluctuations, determines the sediment/water discharge ratio and thus whether stream power exceeds the critical power required to transport the sediment supplied. Incision means that water discharge exceeds the amount needed to carry the available sediment load. Episodic trenching at the fan head has been shown to occur with fluctuations in sediment supply. When the sediment load remained lower than the sediment transport capability, as under sediment-deficient conditions, entrenchment occurred.

Entrenchment depends to some degree on low *sediment availability*. The control of sediment availability on the erosional/depositional trends is demonstrated by flow events that start from an erosional phase, typified by a high water/sediment ratio, followed by a lower water/sediment ratio and change to deposition. During the recessional stage of large foods, when the source terrain still supplies discharge but not much sediment, erosion restarts (Wells and Harvey 1987). Increased runoff periods coincide with erosional trends, whereas periods of high sediment loads with low water discharge do not permit incision and are basically accumulative.

In the long run, low sediment supply will occur as a product of aging (Bull 1977), which causes the progressive diminution of sediment supply after a long term of draining. When the stock of available sediments runs out, entrenchment will start in mountainous catchments and migrate down to the fan. The stage of reduced sediment supply can be considered a sign of *landscape maturity* (Blair and McPherson 1994; Blair 1999b) resulting from an *intrinsic evolution* through time unrelated to external controls (Pepin et al. 2010).

Fan incision may result as well from entrapping sediments. In former glaciated upper source areas, the last glacial maximum formed bedrock barriers that dammed sediments by providing accommodation (Dühnforth et al. 2008). This internal sediment storage significantly delayed the timing of sediment evacuation during the ice-free interglacials or interstadial periods. Without the entrapped sediments, the streamflow became free of bedload and deeply incised the fan heads. A similar effect of sediment retention is produced by the construction of artificial dams. Hillslope vegetation in humid climates similarly holds up transportable sediments and causes relatively load-free floods to entrench the fans. Temporary damming by landslides in the source area can also cause a trench-forming phase in a fan's history. Reduction in sediment input following change to greater or harder bedrock exposed areas in drainage basins will cause a similar sediment deficiency.

The Slope Gradient Factor Stream power depends among others on the gradient. Detachment and sediment transport capacity are power functions of discharge and slope. For any streamflow and sediment load, there is a *threshold slope*, i.e., the slope at which entrenchment will be initiated. Fans that aggrade close to the mountain front will steepen to their critical threshold slope. Their evolutional cycle can be described as aggradation - oversteepening - entrenchment. Oversteepening means exceeding slope stability which is an intrinsic geomorphic threshold (Schumm and Hadley 1957; Schumm 1973; Schumm and Etheridge 1976; Schumm et al. 1987; Weaver 1984). When deposition prevails at the fan head with reduced accretion at the mid- and down-fan areas, the slope at the apex increases beyond threshold and becomes entrenched, and the channel slope becomes gentler. Later backfilling may resume steepening (Schumm 1977). Periods of deposition will thus be interrupted by incision.

The degree of proximity of a fan surface to its critical threshold slope determines whether it will oversteepen and erode in future events. Tectonic uplift of a mountain belt and the upper fan close to the mountain front may contribute to future oversteepening. Fans which are close to their stability threshold need just a small change in external conditions (climate, tectonic) to change trends, for example, from aggradation to incision. Given different background conditions, even adjacent fans may have different thresholds and respond differently to similar external conditions (Schumm et al. 1987).

8.3 Erosional Morphology

Incision As a mountainous catchment starts to erode, incision migrates down to the fan apex, starting the fan entrenchment process that ranges from shallow to deep incision. This is the *top-down driven entrenchment* (Mather et al. 2017) based mainly on climatological- hydrological-sedimentological causes. The uppermost, oldest proximal fan surface is the typical site of *shallow fan-head trenching*. Such entrenchment causes drainage to concentrate along channels that frequently attain a few meters to ~15 m depth and up to ~150 m width. These channels rarely extend beyond the mid-fan zone where they emerge on the fan surface to source a lobe. By trenching, flows cross part of the fan in incised confined conditions attaining greater flow competence (Blair 1999). Shallow entrenched channels may become later completely backfilled (Hooke 1967; Schumm et al. 1987). Proximal fan-head incision or an unincised fan head mark a notable morphologic difference among alluvial fans.

With time, recurrent incision will gradually deepen the channels beyond the point of overtopping and trigger a *deep permanent entrenchment* with typical depths of 20–80 m below the fan-head surface. The final entrenchment stage is deep *through-fan entrenchment*, forming a gorge across the entire length of the fan from the apex to toe, up to a few hundreds of meters wide. The deep entrenchment prevents the supply of any sediment or water to the fan surface, which remains inactive except for soil formation. Sidewalls of neighboring deeply entrenched channels may gradually move toward each other till they merge. With time, the apex elevation declines too (Nicholas and Quine 2007, Fig. 8.1). Lateral erosion in the deeply incised gorges may remove most of the original fan (Timothy and Korup 2007). The fan margins may become entrenched by *interfan streams*, and at the toe a lateral shift of the axial river toward the fan may trigger *headward entrenchment* along shallow gullies, a few meters deep. The gullies grow upslope from the distal fan toe and will gradually be accommodated by the gully geometry and die out.

Continuous fan trenching triggers downslope migration of the locus of deposition. The oldest fan deposits are located near the mountain front, while the younger move down-fan. The evolutional sequence of fan entrenchment – starting from proximal incision and ending in total dissection and lateral erosion that eliminates most of the fan – has been labeled as *fan evolution* (Harvey 2003). Trenching forms

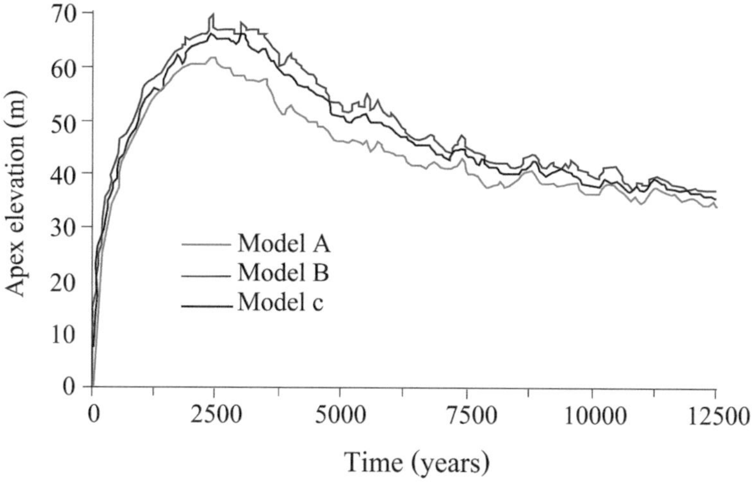

Fig. 8.1 Time series of fan apex elevation based on a lattice-network model of alluvial fan evolution: the fan apex elevation increases reflecting apex aggradation before onset of fan entrenchment (Reproduced from Nicholas and Quine 2007)

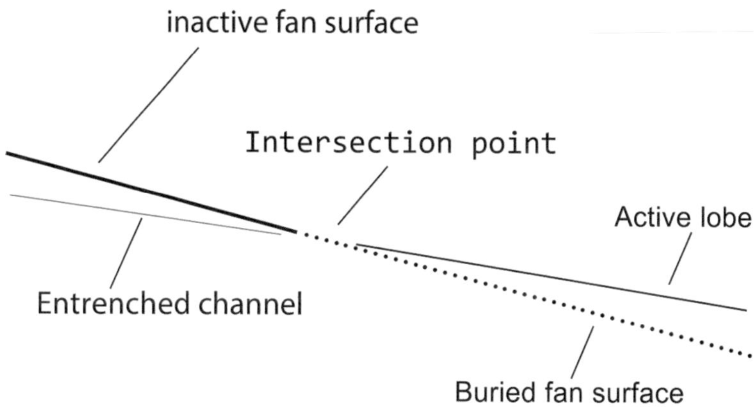

Fig. 8.2 The intersection point environment

more moderate-sloped channels that indicate the trend toward grade. Entrenchment establishes a gentler, stabile channel profile.

Intersection Point The *Intersection point* is the site where an entrenched channel in the apex or mid-fan area emerges onto the fan surface (Hooke 1967; Bowman 1978; Harvey 2011; Fig. 8.2). The intersection point is a mini-apex with the hydraulic status of a flow passing from a narrow exit to a wide dispersal area. Topographic maps may be useful in locating intersection points. Typical contour line irregularities and contour offsets mark the intersection point. A positive dependence between the degree of incision and the magnitude of the contour curvature has been observed.

The intersection point marks a local incision-deposition borderline where the threshold of critical power is crossed and from where down-fan accumulation space becomes available (Weissmann et al. 2002). From the intersection point down-fan, the flow expands, and deposition commences in the form of a small fanlike veneer or *lobe*. The transport capacity from the intersection point down-fan decreases, because of the spreading, even though the flow emerges onto a steeper fan surface. The mean gradient of the entrenched channel is lower than that of the fan surface, and thus an intersection point marks a break in the fan profile. Upslope from an intersection point, the incised channel becomes deeper, and the fan surface is deactivated with soil starting to form.

It is useful to follow the migration of intersection points for revealing trends of deposition or erosion. Aggradation may fill incised channels and cause the intersection points to shift backward toward the apex. During episodes of incision, the intersection points and their associated lobes will shift down-fan. Dating the oldest lobe units below the intersection point constrains the closure age of the entrenchment event and the formation of the intersection point.

When a fan is entrenched and a sequence of intersection points and lobes are formed, a *telescopic fan* is formed. The lobes that prograde out of the intersection points make the longitudinal fan profile highly segmented. The intersection points do not function as a base-level. The channels are neither grading toward nor controlled by the intersection points, which should rather be regarded as *soft knickpoints.*

Coupling or *connectivity* within the fan system (Harvey 2012) describes the degree of the downstream sediment transmission. Alluvial fans as an active sediment storage impedes coupling. Complete connectivity involves the transmission of sediment from the catchment *through* the fan into the downstream receiving basin. This may be accomplished by deep through-fan trenching typified by complete sediment bypass.

References

Blair TC (1999) Sedimentary processes and facies of the waterlaid Anvil Spring Canyon alluvial fan, Death Valley, California. Sedimentology 46:913–940

Blair TC (1999b) Cause of dominance by sheetflood vs. debris-flow processes on two adjoining alluvial fans, Death Valley, California. Sedimentology 46:1015–1028

Blair TC, McPherson JG (1994) Alluvial fans and their natural distinction from rivers, based on morphology, hydraulic processes, sedimentary processes and facies assemblages. J Sediment Res 64:450–489

Bowman D (1978) Determination of intersection points within a telescopic alluvial fan complex. Earth Surf Process 3:265–276

Bull WB (1977) The alluvial fan environment. Prog Phys Geogr 1:222–270

Bull WB (1979) Threshold of critical power in streams. Geol Soc Am Bull 90:453–464

Bull WB (1988) Floods; degradation and aggradation. In: Baker VR, Kochel RC, Patton PC (eds) Flood geomorphology. Wiley, New York, pp 157–165

Dühnforth M, Densmore AL, Ivy-Ochs S, Allen PA (2008) Controls on sediment evacuation from glacially modified and unmodified catchments in the eastern Sierra Nevada, California. Earth Surf Process Landf 33:1602–1613

Harvey AM (2003) The response of dry region alluvial fans to quaternary climatic change. In: Alsharan AS, Wood WW, Goudie AS, Fowler A, Ebdelatif EM (eds) Desertification in the third millennium. Swete & Zeitlinger, Lisse, pp 75–90

Harvey AM (2011) Dryland alluvial fans. In: Thomas DSG (ed) Arid zone geomorphology, process, form and change in drylands, 3rd edn. Wiley, Chichester, pp 333–371

Harvey AM (2012) The coupling status of alluvial fans and debris cones: a review and synthesis. Earth Surf Process Landf 37:64–76

Hooke RL (1967) Processes on arid-region alluvial fans. J Geol 75:438–460

Jones SJ, Frostick LE (2008) Inferring bedload transport from stratigraphic successions: examples from Cenozoic and Pleistocene rivers, south central Pyrenees, Spain. Geol Soc Lond, Spec Publ 296:129–145

Knighton AD (1984) Indices of flow asymmetry in natural streams: definition and performance. J Hydrol 73(1–2):1–19

Laronne JB, Reid I (1993) Very high rates of bedload sediment transport by ephemeral desert rivers. Nature 366:148–150

Leopold LB, Wolman MG, Miller JP (1964) Fluvial processes in geomorphology. Freeman, San Francisco, p 522

Mather AE, Stokes M, Whitfield E (2017) River terraces and alluvial fans: the case for an integrated quaternary fluvial archive. Quat Sci Rev 166:74–90

Nicholas AP, Quine TA (2007) Crossing the divide: representation of channels and processes in reduced-complexity river models at reach and landscape scales. Geomorphology 90(3):318–339

Parker G, Paola C, Whipple KX, Mohrig DC (1998) Alluvial fan formed by channelised fluvial and sheet flow. I. Theory. J Hydraul Eng 124:985–995

Pepin E, Carretier S, Herail G (2010) Erosion dynamics modelling in a coupled catchment-fan system with constant external forcing. Geomorphology 122:78–90

Schumm SA (1973) Geomorphic thresholds and the complex response of drainage systems. In: Morisawa M (ed) Fluvial geomorphology. Publications in Geomorphology, Binghamton, pp 299–310

Schumm SA (1977) The fluvial system. Wiley, New York, p 338

Schumm SA, Etheridge FG (1976) Short course notes on the fluvial system: chapter 7, alluvial fans: Colo. State Univ. Fort Collins, Colo

Schumm SA, Hadley RF (1957) Arroyos and the semiarid cycle of erosion. Am J Sci 255:161–174

Schumm SA, Mosley MP, Weaver WE (1987) Experimental fluvial geomorphology. Wiley, New York, p 413

Timothy RHD, Korup O (2007) Persistent alluvial fanhead trenching resulting from large, infrequent sediment inputs. Earth Surf Process Landf 32:725–742

Weaver WE (1984) Experimental study of alluvial fans

Weissmann GS, Mount JF, Fogg GE (2002) Glacially driven cycles in accumulation space and sequence stratigraphy of a stream-dominated alluvial fan, San Joaquin Valley, California, U.S.A. J Sediment Res 72:240–251

Wells SG, Harvey AM (1987) Sedimentologic and geomorphic variations in storm-generated alluvial fans, Howgill Fells, northwest England. Geol Soc Am Bull 98:182–198

Textural and Facies Characteristics 9

Abstract

Alluvial fans have no overall specific definable texture. Their clast size ranges from boulders to clay and shows a characteristic rapid downcurrent decrease in size. Close similarity in sediment texture from fan head to toe implies that no significant transport-related processes have taken place. Sorting improves down-fan but often remains moderate or still very poor. Normal grading results from a gradual decrease in the sediment supply or from waning flow velocities following floods. Inverse grading is often observed with stratigraphic thickening and corresponds to a prograding gravel front. The overall thickness of stratigraphic fans units ranges from several meters to a great thickness of a few kilometers. The thickest part of a fan usually leans against the mountain front.

Keywords

Fanglomerate · Normal grading · Inverse grading · Coarsening upward · Progradation · Retrogradation · Mountain-front embayment · Channel fill · Interfingering · Cyclothems · Couplets · Facies associations

9.1 Grain Size

Fans range in clast size from boulders to clay and have no overall specific definable grain size. However, many fan sediments are typically coarse and consist of pebble- to meter-sized boulders gathered under the term *fanglomerate*. The fan sediments may, however, also reside on the fine and smooth side of the textural spectrum and consist entirely of sand, silt, and clay. Because of the proximity of alluvial fans to their sourcing area and the relative short distance of transportation, textural immaturity is typical of fan deposits.

The proximal fan zone shows an immature, coarse texture, often with boulders several m in size. The ability of a fan to transport the gravel fraction diminishes

© Springer Nature B.V. 2019
D. Bowman, *Principles of Alluvial Fan Morphology*,
https://doi.org/10.1007/978-94-024-1558-2_9

exponentially down-fan (Stock et al. 2007) with the largest concentrations of fines to be found in the lower fan zone. Fining down-fan (Fig. 9.1) is often higher by orders of magnitudes than appears possible for abrasion alone and is thus evidently not a result of clast attrition but of selective deposition due to rapidly declining transporting capacity (Paola et al.1992). Nevertheless, lack of downcurrent bedload fining along large sections of the upper and middle fan channels where slope declines has also been reported (Stock 2013). Close similarity in sediment texture from fan head to the toe has been reported as well, implying that no significant transport-related processes have occurred. Clast sizes that are present at the apex and are missing at the toe must have been deposited on the fan. Using the clast size, empirical functions can be applied to calculate the palaeovelocities required to move the gravel.

Fines on fans do not always originate solely from the source basin but may also be wind-imported to the fans from deflating areas in the near region. Part of the fine material is winnowed by water along the fan surface and accumulates below the surface to form the vesicular A layer of the soil or fills interstices in the gravel. Part of fines may be winnowed away, bypass the fan, leaving a surface roughened with the coarser clastics.

When channels change down-fan from deeply entrenched to shallow flows, fining is not self-similar, i.e., the standard deviation (σ) of the grain size distribution does not decay at a similar exponential rate as the mean size (D) and thus the coefficient of variation $Cv = \sigma/D$ does not remain constant. When the standard deviation of the grain size decreases faster than the mean size, a downstream sorting adjustment is indicated (Miller et al. 2014). Sorting improves down-fan but often remains

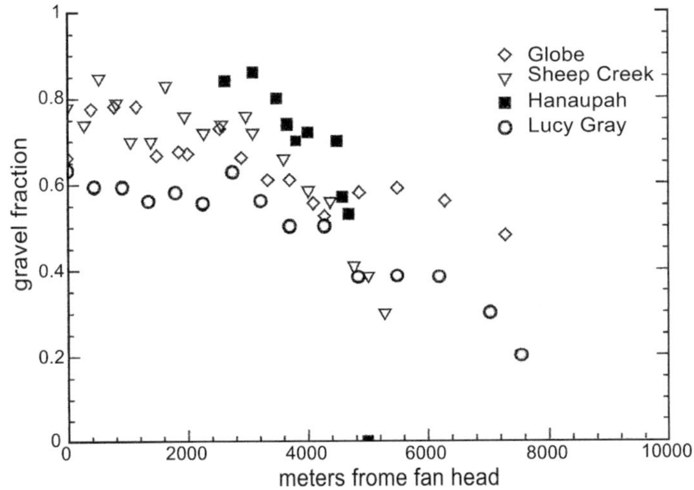

Fig. 9.1 Reduction of bed area gravel fraction down-fan based on data of four fans of the Western USA: Globe, Sheep Creek, Hanaupah, and Lucy Gray. Reduction is partly exponential. (Following Stock et al. 2007)

moderate or very poor because of short distance of transport. Roundness may increase from the proximal to the distal areas, changing from angular/subangular or subrounded up to very well rounded, including boulders.

Fining upward or *normal grading* which is seen in clast-supported fan deposits (Fig. 9.2) indicates a gradual decrease in sediment supply or a waning of flow velocities following floods. It may as well testify to the change from channel transport to unconfined sheetfloods. Fining upward may also reflect the downcutting and subduing topographic relief at the source area with time or occur following a slowdown of tectonic uplift. *Coarsening upward* or *inverse grading* is the stratigraphic upward increase in the maximum particle size, often observed with stratigraphic thickening and exposure of large boulders. Coarsening upward often corresponds to a *progradational stage*, i.e., when coarse lobes from the apical zone prograde downward and bury parts of the mid-fan and the down-fan areas. Coarsening upward to become a massive conglomerate has been suggested by Kumar et al. (2007) to indicate a prograding gravel front in response to a rapidly subsiding basin. Generally, supply of thick coarse clasts is often related to tectonic rejuvenation, including thrusting, which increases relief and steepness. However, upward coarsening is not an incontroversial evidence of tectonically induced erosion. Climatically controlled events can produce sedimentary signatures similar to those created by tectonism. Coarser stratigraphy may indicate periods of increased rainfall and subsequent increased discharge with greater flow velocities or may reflect change in the outcrop lithologies of the source area or in the availability of pre-existing coarse sediments as glacial till and their opening to mobilization (Dec 1992).

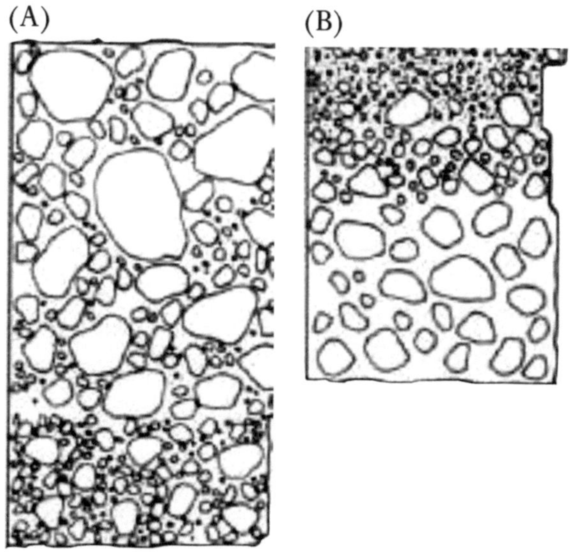

Fig. 9.2 Typical examples of (**a**) fining and (**b**) coarsening upward in conglomerates. (Modified after Dec 1992)

9.2 Facies

Fan facies is distinguished on the basis of the following parameters (modified from Miall 1978): grain size, matrix composition, sorting and packing, clast or matrix support, stratification or massiveness, upward fining or coarsening, and clast imbrication. The sum of these facies criteria allows distinction of a proximal, medial, and distal subenvironment (Blair 1999; Figs. 1.8 and 9.3). We can display the following typical zonal features (Fig. 9.4):

1. *The proximal fan facies* The proximal fan segment consists of sediments stacked against a fault face. Proximal sediments are the first to be deposited and often form the oldest fan segment. The sediments accumulate as a massive, structureless pile or with crude stratification, characteristically disorganized, poorly to extremely poorly sorted, and without preferred orientation. The proximal facies is dominated by boulder to pebble clast sizes with occasional outsized boulders and an open-framework clast-supported mass of gravel. Postdepositional sand or sandy granules may appear as an interclastic matrix, formed by entrapment and

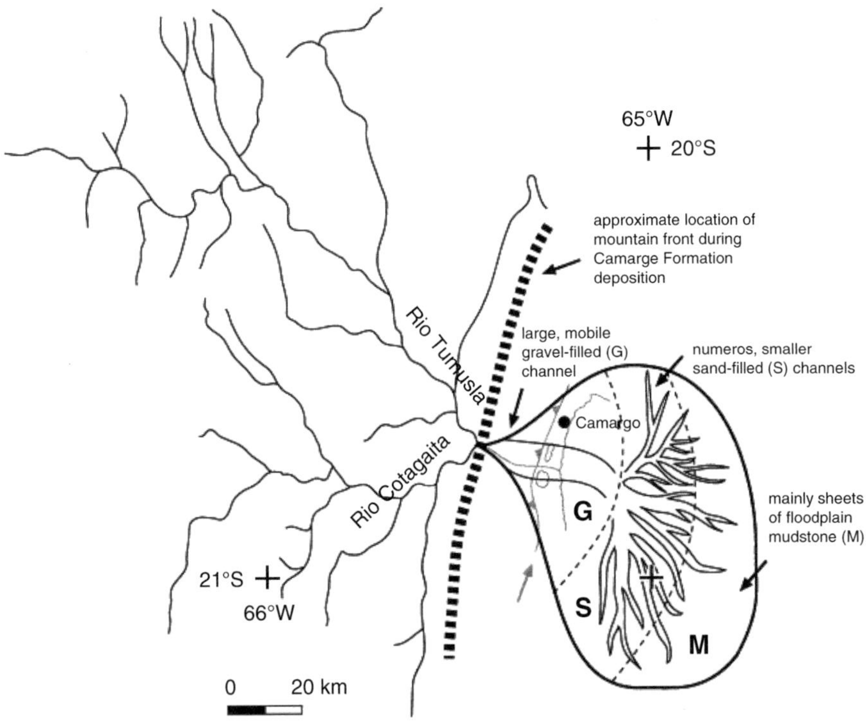

Fig. 9.3 The main fan environments (G, S, M) displayed in a depositional model of a megafan and its sourcing drainage basins at the Eastern Cordillera of the central Andean fold belt. (Modified after Horton and DeCelles 2001)

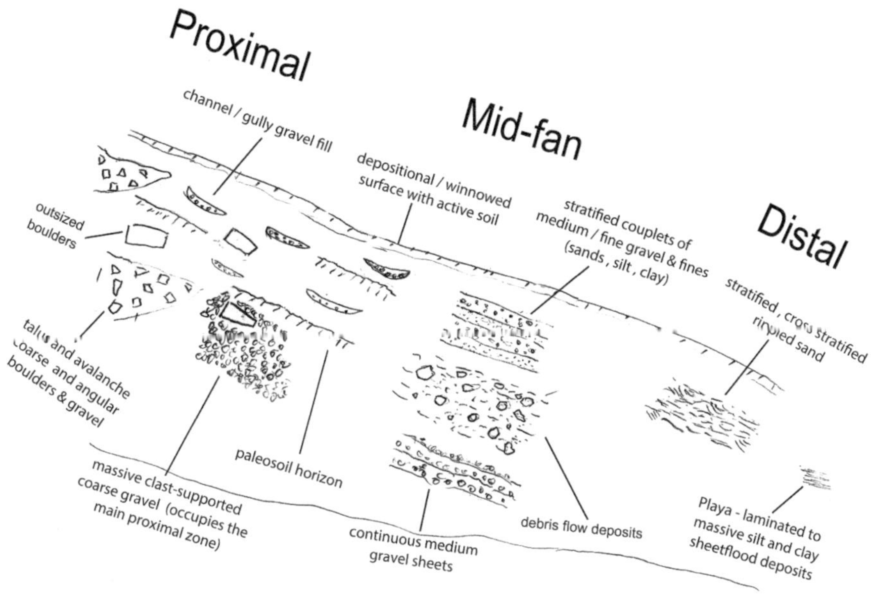

Fig. 9.4 Schematic longitudinal fan section with typical stratigraphic features. Interfingering and lateral continuity are not portrayed. In logs the lateral stratigraphic changes show vertically

infiltration during waning flood stages. Fine-grained overbank deposits are usually not typical of the proximal zone.

The signature of the proximal zone reflects more often erosion and reworking rather than primary deposition. This erosional stage is typified by scour and refill of incised channels set into a nonchannelized facies and filled by massive pebbly to bouldery lag conglomerates, often as steep foresets or trough cross-bedding (Mack and Leeder 1999). The channels are trough-shaped in cross section, about 7–10 m deep, and can be observed in exposures cut at right angle to the flow direction. The coarse-grained texture of the fill, its upward fining, and the wedge-shaped geometry are typical of high-energy channels. Some fans may become filled backward into their feeder and form a *mountain-front embayment* (Figs. 1.8, 1.9, and 9.5c).

2. *The mid-fan facies* Mid-fan sediments are intermediate between the coarseness of the apex and the finer fractions down-fan. The mid-fan displays sedimentary alternations of coarse and fine deposits and an increase in the finer-grained sheet deposits which gradually take over. Soil horizons testify to periods of inactivity and abandonment of wide portions of the mid-fan surface.
3. *The distal fan facies* The distal areas display a transition from channel flow to shallow sheet flows. The slowing of the flow in the distal areas is caused by the high hydraulic roughness, the higher density of vegetation, and the low slope gradients. Distal flats typically show an overall well-stratified, sheetlike, often laminated, structure. Thin fills of shallow channels are granules, sand, silt, and clay. The flats often display alternating coarser and finer deposits defined *couplets*, several tens of

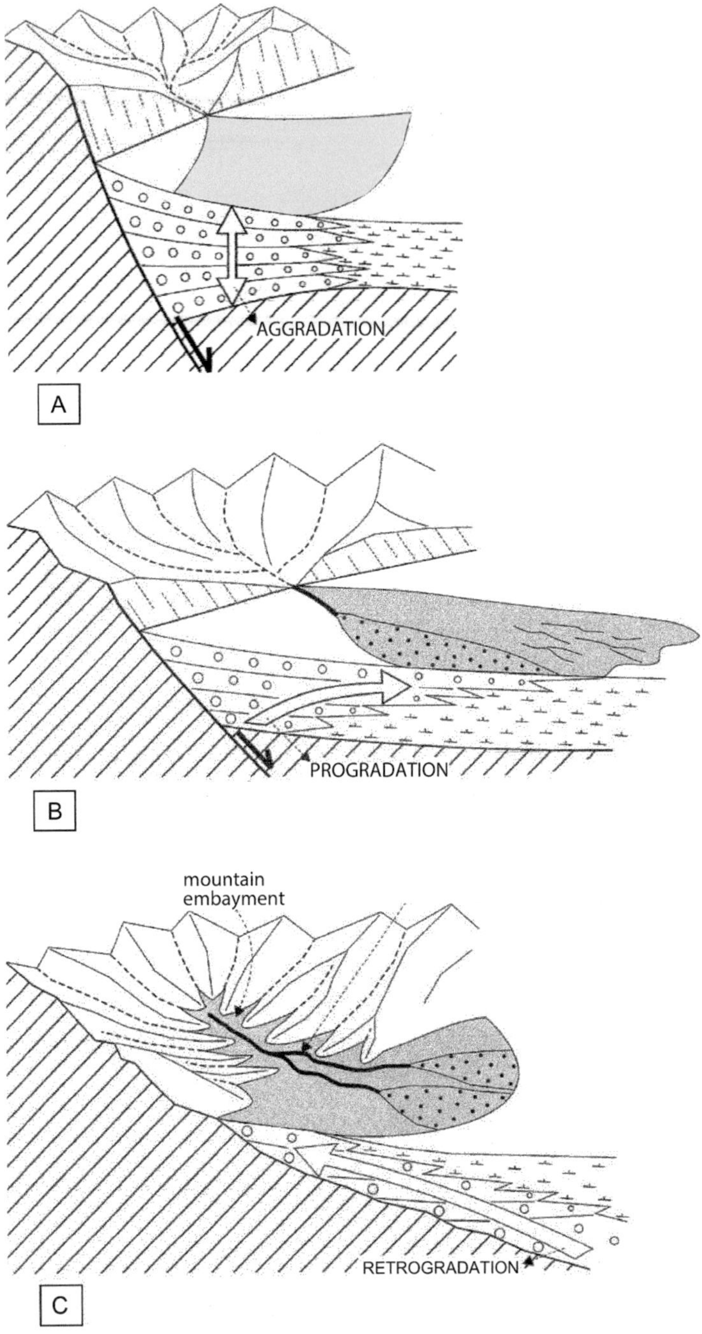

Fig. 9.5 Stratigraphic patterns of fans when accommodation: (**a**) is balanced by sediment supply – *vertical stacking*; (**b**) when formation of accommodation space lags behind sediment discharge – *progradation*; and (**c**) when low transportation capacity dominates without available accommodation space causing embayment fill – *retrogradation*. (Modified after Viseras et al. 2003)

centimeters thick (Blair 2000). In spite of the tendency of sediments to become finer down-fan, boulders 1–2 m in diameter may surprisingly still be prominent at the lower margins of the fans. Even extreme floods do not attain the calculated critical threshold required to mobilize such considerable size. To continue mobilizing such boulders, an *undermining*, destabilizing mechanism had been suggested for rolling the boulders without exceedance of the required critical threshold (Storz-Peretz et al. 2011).

Alluvial fans often overlie former lake sediments that are the basin fill. Thickness usually diminishes down-fan (Fig. 9.5a, b) in harmony with the diminishing maximum particle size. A lake level rise onto a fan will trigger a retrograding alluvial trend and an upward migration of the fan deposits (Schumm 1993; Fig. 9.5c). Prograding fans may move over adjacent fine playa sediments or over deposits of the axial river (Fig. 9.5b). The fan stratigraphy displayed in vertical sedimentary profiles is the most efficient means of establishing trends of retrogradation, progradation, stacking, or entrenchment. Cyclic stratigraphic units – *cyclothems* – record clastic and pedogenic cyclicity indicating periods of fan surface stability and soil formation followed by lateral shifting of channels. However, stratigraphic units in alluvial fans need not be synchronous and are thus not useful for wide lateral extrapolation (Swanson-Hysell and Barbeau 2007). Sampling strategies should minimize lateral and down-dip correlation distances.

Because of fan avulsion, there is no single site on a fan where its evolutional history can be fully recorded from stratigraphy, hence fan studies should take advantage of all large-scale outcrops along the fan head as well as along banks of entrenched channels and gullies for documenting the textural trends. *Facies associations* which comprise units of similar facies characteristics should be described (Hornung et al. 2010), and the stratal architecture should be recorded by line drawing of outcrops, both parallel and transverse to the flow direction.

Sampling along transect lines and outcrops and measuring structural and morphological dips is recommended as well as sampling sediments from cores to assess porosity characteristics, sedimentary structures, and paleomagnetic and hydraulic properties. The fan stratigraphy can be further investigated by interpretating vertical electrical soundings and by examining subsurface correlation of boreholes. Down-fan and across-fan comparison of sections may disclose tectonic or climatic signals (Suresh et al. 2007).

References

Blair TC (1999) Sedimentary processes and facies of the waterlaid Anvil Spring Canyon alluvial fan, Death Valley, California. Sedimentology 46:913–940

Blair TC (2000) Sedimentology and progressive tectonic unconformities of the sheetflood-dominated Hell's Gate alluvial fan, Death Valley California. Sediment Geol 132:233–262

Dec T (1992) Textural characteristics and interpretation of second-cycle, debris-flow-dominated alluvial fans (Devonian of Northern Scotland). Sediment Geol 77:269–296

Hornung J, Pflanz D, Hechler A, Beer A, Hinderer M, Maisch M, Bieg U (2010) 3-D architecture, depositional patterns and climate triggered sediment fluxes of an alpine alluvial fan (Samedan, Switzerland). Geomorphology 115:202–214

Horton BK, DeCelles PG (2001) Modern and ancient fluvial megafans in the foreland basin system of the central Andes, southern Bolivia: implications for drainage network evolution in foldthrust belts. Basin Res 13(1):43–63

Kumar R, Suresh-Satish N, Sangode J, Kumaravel V (2007) Evolution of the quaternary alluvial fan system in the Himalayan foreland basin: implications for tectonic and climatic decoupling. Quat Int 159:6–20

Mack GH, Leeder MR (1999) Climatic and tectonic controls on alluvial-fan and axial fluvial sedimentation in the Plio-Pleistocene Palomas half graben, southern Rio Grande rift. J Sediment Res 69:635–652

Miall AD (1978) Lithofacies types and vertical profiles models in braided river deposits: a summary. In: Miall AD (ed) Fluvial sedimentology. Can Soc Petrol Geol Mem 5: 597–604

Miller KL, Reitz MD, Jerolmack DJ (2014) Generalized sorting profile of alluvial fans. Geophys Res Lett 41:7191–7199. https://doi.org/10.1002/2014GL060991

Paola C, Hellert PL, Angevinet CL (1992) The large-scale dynamics of grain size variation in alluvial basins 1: theory. Basin Res 4:73–90

Schumm SA (1993) River response to base level change: implications for sequence stratigraphy. J Geol 10:279–294

Stock JD (2013) Waters divided: a history of alluvial fan research and a view of its future. In: Shroder JF, Wohl E (eds) Treatise on geomorphology, Fluvial Geomorphology, vol 9. Academic, San Diego, pp 413–458

Stock JD, Schmidt KM, Miller DM (2007) Controls on alluvial fan long-profiles. Geol Soc Am Bull 120:619–640

Storz-Peretz Y, Bowman D, Laronne JB, Svoray T (2011) Rapid incision of a small, coarse and steep fan-delta in response to base-level fall: the case of Nahal Qedem, the Dead Sea, Israel. Earth Surf Process Landf 36:467–480

Suresh N, Bagati TN, Kumar R, Thakur VC (2007) Evolution of quaternary alluvial fans and terraces in the intramontane Pinjaur Dun, Sub-Himalaya, NW India: interaction between tectonics and climate change. Sedimentology 54:809–833

Swanson-Hysell N, Barbeau DL Jr (2007) The diachroneity of alluvial-fan lithostratigraphy? A test case from southeastern Ebro basin magnetostratigraphy. Earth Planet Sci Lett 262(3):343–362

Viseras C, Calvache ML, Soria JM, Fernández J (2003) Differential features of alluvial fans controlled by tectonic or eustatic accommodation space. Examples from the Betic Cordillera, Spain. Geomorphology 50(1–3):181–202

Pedogenic Processes

10

Abstract

Soils in the fan's stratigraphic column testify to a period of surface exposure and stability during the depositional phase. This happens when deep entrenchment leaves parts of the fan surface inactive for longer periods. The paleosol thus marks a depositional hiatus, i.e., an inactive period within the depositional phase. In extremely arid to semiarid climates, the typical product is the *reg soil* marked by a vesicular, fine-grained Av horizon and a B horizon enriched in carbonate, gypsum, or salts. *Gravel shattering* by salts is found both on the fan surface and in the B and C horizons.

The abandoned fan surfaces become a desert pavement composed of closely packed, angular rock fragments, making a mosaic mantle that protects the underlying soil from removal. Desert pavements persist as long as the stone coverage remains above 60% and provide a relative-age estimator that progresses concurrently with the soil formation. With time, the gravel on the fan surface becomes coated with *desert varnish* which consists of layers of hydrous ferric manganese oxide, iron oxides, clay minerals, and trace elements. The Fe oxide typically amounts to less than 1%. Its percentage of cover and the darkness of its varnish increase systematically with age. Carbonate encrustation is another characteristic feature of semiarid fan surfaces. The higher and older the fan surfaces are, the better preserved the carbonate encrustations become.

Keywords

Paleosol · Depositional hiatus · Reg soil · Petrocalcic "K" horizon · Gravel shattering · Salt weathering · Desert varnish · Soil rubification · Desert pavement

© Springer Nature B.V. 2019
D. Bowman, *Principles of Alluvial Fan Morphology*,
https://doi.org/10.1007/978-94-024-1558-2_10

Soil in the fan stratigraphic column is evident of a long period of surface exposure with little or no depositional activity, negligible input of runoff, and thus resistance to erosion. Such soil of the past, exposed in the stratigraphy, is named *paleosol* and indicates deep entrenchment or channel migration which left parts of the fan surface inactive for long periods. The paleosol thus designates a *depositional hiatus*, i.e., an inactive period within a depositional phase. The absence of paleosols in a stratigraphic column suggests a relatively rapid and continuous aggradation without fan entrenchment that would have enabled fan surface abandonment. Soil development over an entire fan surface without entrenchment may reflect loss of the catchment sourcing through stream capture (Stokes et al. 2007).

10.1 Development of Reg Soils

Where climates are extremely arid (<80 mm/y) to semiarid (150–200 mm/y), inactive fan surfaces develop *reg soils* which are as a rule saline or gypsiferous, typified by oxidizing conditions and by very small amounts of organic matter. The reg soil is formed from the underlying parent material combined with import of eolian dust and may reach a depth of 30–100 cm and in some Middle to Late Pleistocene soils even more (Gerson 1981; Amit and Gerson 1986).

Pedogenic processes begin shortly after the fan surface became inactive (Yaalon 1971; Amit et al. 1993). A vesicular, loamy, fine-grained horizon (McFadden et al. 1998) called "Av" horizon is formed mostly through accumulation of eolian dust which requires the bar and swale micromorphology of the fan as a rough surface to be entrapped with the airborne salts and be protected from deflation and raindrop impact. The thickness of the Av horizon usually ranges from a few mm to 20 cm. The pH ranges most commonly from 7.8 to 8.4. The texture ranges from sandy clay to silt or sandy loam. Organic matter content of the horizon ranges from zero to 0.1%.

Reg profiles show the silty A horizon and a gravelly C horizon within 1000–2000 years. On alluvial surfaces 7000–14,000 years old, the C horizon occupies most of the soil profile. Clay particles are partly flushed by percolating water from the Av into the Bt argillic horizon. Dust continues to accumulate, thickening the B horizon, sealing the soil, and limiting the rate of water penetration. The B horizon becomes with time enriched by carbonate, gypsum, or salts. Carbonate grows as coatings around clasts and cementing matrix between grains. Eventually the B horizon becomes entirely or almost gravel-free. When mature, the *reg soil* displays a well-defined vesicular A horizon and a gravel-poor, reddish B horizon with accumulation of gypsum, soluble salts, or carbonate coating. Shattering of gravel clasts by crystallization of salts exhibits a high density of cracks.

The development of soil indicates the initial fossilization of the fan surface and may serve for relative-age dating and for correlating fan surfaces (McFadden et al. 1989). The time required for the formation of fan soils needs to be of sufficient duration to produce pedogenesis. Short stable periods may be insufficient to attain a significant soil evolution. It takes some 25 ky to produce sufficiently clay concentrations and enable differentiation of horizons. The proximal fan surface, which is the oldest and the first to become entrenched, usually shows the most developed soil profile.

10.2 Calcretization and Salt Weathering

Carbonate encrustation in the soil profile develops when calcium, derived from external carbonate sources as windblown dust or from calcium-bearing minerals, is leached from the upper soil horizon downward (Gile et al. 1966). Such calcrete development is a characteristic feature of modern as well as Pleistocene semiarid fan surfaces (Alonso-Zarza et al. 1998). As the growth of coalescing carbonate nodules progresses, calcite precipitation extends throughout the uppermost part of the soil profile, and laminar carbonate encrustation on clasts grows from patchy to massive. Calcretes reduce soil porosity, resulting in reduced infiltration and increased volume of surface runoff (Lattman 1973). The strongly impregnated uppermost part of the soil horizon is the *petrocalcic "K" horizon*. The best developed carbonate stages are no. III–V of Gile et al. (1966) and of Machette (1985).

A mature calcrete profile with a well-developed upper "K" horizon may be over 2–3 m thick. Such massive soil and crust development suggests a considerable period of fan stability gained following a deep entrenchment into the upper fan area (Alonso-Zarza et al. 1998). When calcrete encasement acts as a caprock and sufficiently strengthens the surface, fans may reveal negligible surface incision, delayed entrenchment, and headward erosion and thus well preserve the higher and older fan outliers. Knickpoints may form by channels incising through the indurated calcic horizons.

Gravel shattering by salts found both on the fan surface and in the B and C horizons is common in desert reg soils (Yaalon 1970; Amit and Gerson 1986; Amit et al. 1993). The depth of the saline layer corresponds roughly to the depth of maximum water penetration. Gravel clasts are shattered by salt crystallization, salt hydration, and thermal expansion. The disintegration is affected by the salt crystal growth which causes expansion and requires space. Pressure that gradually accumulates on walls of cracks causes them to shatter when exceeding the clast's mechanical strength.

The overall process of shattering is not linear but self-terminating, i.e., the rate is rapid in the first 10,000 years, but with time the rate slows down and stabilizes (Yaalon 1971). In the first stages of shattering with very small amounts of salt, the shattered gravel reached 15–20% (Amit and Gerson 1986). Salinity becomes definitely higher in reg soils older than some 2000–3000 years (Amit et al. 1996). In reg soils dated of Late Pleistocene (14,000 years), the salinity was found to be 0.27–16 mmho/cm with a percentage of 20–70% shattered gravel which still maintained their original shape. In a mature reg soil of Middle Pleistocene, about 500,000 years old, with high salt amounts of 62 mmho/cm, the percentage of the shattered gravel amounted to 85%. The shattering process can be split into stages with the shattering percentage used as a relative-age indicator.

10.3 Varnish and Soil Rubification

Gravel that covers the fan soil in arid zones is coated by *desert varnish* which consists of layers of hydrous ferric manganese oxide, iron oxides, clay minerals, and trace elements. Varnish is brownish-black when rich in manganese or reddish-brown when iron-rich and has a low albedo. The varnish surfaces are often associated with small bacterial colonies.

The percentage of cover and the darkness of the varnish increases systematically with age and thus provides a relative timescale for fan surface dating (Dorn and Oberlader 1982; McFadden et al. 1989). Varnish develops relatively fast, i.e., in an order of a few thousand years, and thus also characterizes young reg soils. In a time range of 10^3–10^6 years, inactive fan surfaces gradually become darkened by desert varnish. Clasts deposited less than 2000 years before present show only minimal varnishing (Dorn and Oberlader 1982; McFadden et al. 1987). Dark desert varnish goes with mature soil profiles, thick Av horizon, a developed Bt horizon, and an advanced stage of carbonate Bk horizon.

Quaternary soils become redder as they develop. Fan surface stability enhances iron oxide accumulation and accompanying reddening. This process, known as *soil rubification*, is caused by leaching of iron under oxidizing conditions from iron-bearing minerals in the soil's parent material (Birkeland 1974). Fe-oxide hematite is the predominant Fe-oxide mineral to accumulate in the soils (Mcfadden 1985) in concentrations less than 1%. Redness of the soils is linearly correlated with hematite content (Hurst 1977) and is time-dependent. The Munsell soil color chart and other scales that categorize redness are not able always to discriminate between the different reddening stages.

Soil redness and the extractable secondary iron oxide concentration from the B horizon show a consistent and statistically significant gradual increase from lower and younger fan zones through the higher fan terraces up to the highest proximal, oldest fan surfaces (Pope and Wilkinson 2005; Bowman 1982; White and Walden 1997). When lack of organic matter hampers conventional radiocarbon dating of soils on geomorphological surfaces, the rate of pedogenic Fe-oxide enrichment becomes an important proxy for relative dating. Multispectral remotely sensed data based on the distinct spectral reflectance properties of soils imparted by different Fe-oxide concentrations enables soil mapping on fan surfaces (Ferrier and Pope 2012; White 1993; White et al. 1992).

10.4 Desert Pavement

Desert pavement is the smooth, stony mosaic mantle overlying the reg soil (McFadden et al. 1987). A well-developed desert pavement consists of one-to two-particle-thick layers of closely packed, angular to subrounded rock fragments, making a continuous, flat stony veneer. Pavements persist as long as pebble density remains above 60%. The clasts protect the underlying soils from removal by runoff and by wind and cause a distinct decrease of infiltration. With increasing pavement age, the gravel coating becomes darker, and the gravel becomes more angular through cracking.

The development of smooth pavement surfaces results from the progressive breakdown of alluvial clasts following intense salt splitting and granular disintegration. Older pavements consist entirely of small angular gravel. Soil development involving thicknesses, horizon development, redness, and carbonate or gypsum accumulation progresses concurrently with pavement formation which thus provides a relative-age estimator. The pavement as an age index can be based on clast size, on clast angularity, on fracturing, and on the darkness of the surface hue. Amit and Gerson (1986) identified immature pavements on surfaces younger than 14 ka, whereas 83–130 ky were needed for a smooth pavement to form (Reheis et al. 1992; Amit et al. 1993). Pavements develop linearly (Al-Farraj and Harvey 2000) and may be useful to separate Early, Middle, or Late Holocene periods.

For covering the pedogenic aspect in alluvial fan studies, key soil properties should be described through excavating test pits at distinct patches on the fan. The descriptions should include the pavement particle size and clast angularity, the soil thickness, the B horizon including the characteristic soil colors, the carbonate accumulation stage using the Gile et al. (1966) classification scheme modified by Machette (1985), the C horizon texture as representative of the parent material, and the laboratory analysis of iron oxide chemistry. The soil surveys should serve correlation of geomorphic fan units and of the subsurface stratigraphy.

Spectral analysis data can be used for mapping pavements based on relating specific absorption features to ground-surveyed parameters (Crouvi et al. 2006). Fan surface reflectance is controlled by two main surficial features, which are used as relative-age criteria: (a) the degree of desert pavement development shown by the percentage of the gravel coverage (as the desert pavement develops, the absorption feature of the dominant gravels is increased) and (b) rock coating that exhibits spectral reflectance lowering which increases with exposure time. It is possible to map the percentage of the surface gravel coverage on the alluvial fan from the spectral absorption values at 95% confidence interval with ±15% gravel coverage error.

References

Al-Farraj A, Harvey AM (2000) Desert pavement characteristics on wadi terrace and alluvial fan surfaces: Wadi Al-Bih, UAE and Oman. Geomorphology 35:279–297

Alonso-Zarza AM, Silva PG, Goy JL, Zazo C (1998) Fan-surface dynamics and biogenic calcrete development: interactions during ultimate phases of fan evolution in the semi-arid SE Spain (Murcia). Geomorphology 24:147–167

Amit R, Gerson R (1986) The evolution of Holocene reg (gravelly) soils in the deserts – an example from the Dead Sea region. Catena 13:59–79

Amit R, Gerson R, Yaalon DH (1993) Stages and rate of gravel shattering process by salts in desert reg soil. Geoderma 57:295–324

Amit R, Harrison JBJ, Enzel Y, Porat N (1996) Soils as a tool for estimating ages of Quaternary fault scarps in a hyper-arid environment – the southern Arava valley, the Dead Sea Rift, Israel. Catena 28:21–45

Birkeland PW (1974) Pedology, weathering and geomorphological research. Oxford University Press, Oxford
Bowman D (1982) Iron coating in recent terrace sequences under extremely arid conditions. Catena 9:353–359
Crouvi O, Ben-Dor E, Beyth M, Avigad D, Amit R (2006) Quantitative mapping of arid alluvial fan surfaces using field spectrometer and hyperspectral remote sensing. Remote Sens Environ 104:103–117
Dorn RI, Oberlader TM (1982) Rock varnish. Prog Phys Geogr 6:317–367
Ferrier G, Pope RJ (2012) Quantitative mapping of alluvial fan evolution using ground-based reflectance spectroscopy. Geomorphology 175–176:14–24
Gerson R (1981) Geomorphic aspects of the Eilat Mountains. In: Dan J, Gerson R, Koyumdjisky H, Yaalon D (eds) Aridic soils of Israel. International conference on aridic soils, pp 279–296
Gile LH, Peterson FF, Grossman RB (1966) Morphological and genetic sequence of carbonate accumulation in desert soils. Soil Sci 101:347–360
Hurst VJ (1977) Visual estimation of iron in saprolite. Geol Soc Am Bull 88(2):174–176
Lattman LH (1973) Calcium carbonate cementation of alluvial fans in southern Nevada. Geol Soc Am Bull 84:3013–3028
Machette MN (1985) Calcic soils of the southwestern United States. Geol Soc Am Spec Pap 203:1–22
McFadden LD (1985) Changes in the content and composition of pedogenic iron hydroxides in a chronosequence of soils in Southern California. Quat Res 23:189–204
McFadden LD, Wells SG, Jercinovich MJ (1987) Influences of eolian and pedogenic processes on the origin and evolution of desert pavements. Geology 15:504–508
McFadden LD, Ritter JB, Wells SG (1989) Use of multi-parameter relative-age methods for age estimation and correlation of alluvial fan surfaces on a desert piedmont, eastern Mojave Desert, California. Quat Res 32:276–290
McFadden LD, McDonald EV, Wells SG, Anderson K, Quade J, Forman SL (1998) The vesicular layer and carbonate collars of desert soils and pavements: formation, age and relation to climate change. Geomorphology 24:101–145
Pope RJJ, Wilkinson KN (2005) Reconciling the roles of climate and tectonics in Late Quaternary fan development on the Spartan piedmont, Greece. Geological Society London. Special Publications 251 pp 133–152
Reheis MC, Sowers JM, Taylor EM, McFadden LD, Harden JW (1992) Morphology and genesis of carbonate soils on the Kyle Canyon fan, Nevada, USA. Geoderma 52(3–4):303–342
Stokes M, Nash DJ, Harvey AM (2007) Calcrete 'fossilisation' of alluvial fans in SE Spain: the roles of groundwater, pedogenic processes and fan dynamics in calcrete development. Geomorphology 85:63–84
White K (1993) Image processing of Thematic Mapper data for discriminating piedmont surficial materials in the Tunisian Southern Atlas. Int J Remote Sens 14:961–977
White K, Walden J (1997) The rate of iron oxide enrichment in arid zone alluvial fan soils, Tunisian southern atlas, measured by mineral magnetic techniques. Catena 30:215–227
White K, Walden J, Rollin EM (1992) Remote sensing of pedogenic iron oxides using Landsat Thematic Mapper data of southern Tunisia. Proc. of 18th Annu. Conf. of The Remote Sensing Soc., Dundee, pp 179–187
Yaalon DH (1970) Parallel stone cracking, a weathering process on desert surfaces. Geol Inst Bucharest: Technol Econ Bull 18:107–111
Yaalon DH (1971) Soil forming processes in time and space. In: Yaalon DH (ed) Paleopedology origin, nature and dating of paleosols. Inter. Soc. Soil Sci. and Israel Universities, pp 29–39

The Tectonic Control

11

Abstract

Regional tectonics determines the location of alluvial fans by creating sharp topographic differences. Tectonic uplifts enhance bedrock erosion by supporting the capacity to detach and transport clasts. To ensure a continuous fan growth, a long-time trend of uplift or sinking is required.

Being deposited atop a tectonically mobile zone, the alluvial fans too undergo themselves tectonic deformation. Strike-slip faults transfer their motion to the overlying fans, causing the fan segments to be shifted laterally. The amount of such horizontal slips can be determined by the offset of geomorphic markers. Any morphologic surface direction that is at variance with the radial channel pattern and significantly disturbs fan concentricity might be due to tectonic deformation.

Mountain-front sinuosity estimates the irregularity of mountain fronts and provides an evaluation of the balance between stream erosion and weathering relative to tectonic uplift. The mountain-front faulting controls the form and the volume of the accommodation space. When the rate of delivery from the hinterland exceeds the rate of creation of space to store the sediments, fan gravel progradation is enhanced. Coarsening upward points to gravel progradation usually related to tectonic uplift events.

Keywords

Sinuosity index · Mountain front embayment · Accommodation space · Stacking · Progradation · Dead faults · Quantitative morphologic dating · Stratigraphic separation

© Springer Nature B.V. 2019
D. Bowman, *Principles of Alluvial Fan Morphology*,
https://doi.org/10.1007/978-94-024-1558-2_11

11.1 Faulting of Alluvial Fans

The foremost external factors that control the formation and evolution of alluvial fans are *climate*, *base-level*, and *tectonics*. These make alluvial fans recorder of past tectonics, of climates, and of base-level fluctuations. Regional tectonics in the form of faulting, folding, or thrusting determines the location of alluvial fans by creating sharp topographic differences between adjacent terrains (Harvey 1990; Burbank and Anderson 2012). Fans are characteristic features along border lines of active tectonics that build the relief necessary to feed the fans (Bull 1964; Nichols 1987). Long-lasting trends of uplift or sinking are important for maintaining the altitudinal differences which ensure continuous fan sourcing over long periods. In-between tectonic events, climatic changes may determine rates of aggradation or incision.

Being piles of sediments deposited atop tectonically mobile zones, alluvial fans often become deformed, bearing clear signs of surface rupture (Bull 1996). A fan may become tilted following tilting of the block below which initiates progradation and elongation in the direction of the tilt. By examining malformed topographic contours, an estimation of the tilt can be achieved (Pinter and Keller 1995). Any direction of the surface fan morphology that is at variance with the radial channel pattern and any disturbed fan concentricity might be due to tectonic deformation.

The radial braiding pattern on surfaces of fans serves as a useful background for detecting drainage misalignments following tectonic offsets. Identification of faults on fans is as well often based on linear fault scarps, accompanied by a shallow morphological saddle, crossing the fan surface normal to the drainage network or on a close to 90° change in the direction of a channel. Other displaced or ruptured geomorphic features on fans may include deflected terrace risers, linear fan edges, shutter and push-up ridges, and graben structures with sharp scarps. Fans may also show back-tilting of the fan surface in the upfan direction and open tension fissures filled with collapsed sediments (Amit et al. 2002). Subsurface displacements are observed in trenches, including rotated clasts and layers displaced by *dead faults*, i.e., faults that die out below and do not rupture the surface.

Fault scarps on fans are typically several decimeters to a few meters high, veneered by an apron that becomes less steep with age. The aprons have been extensively used to determine the age of ruptures. Quantitative morphologic dating techniques use diffusion equations to estimate the age of fault scarps out of the scarp slope (Bucknam and Anderson 1979; Hanks and Wallace 1985; Hanks 2000). It is assumed that the scarps were formed during a single event and that linear mass diffusivity is a reasonable approximation for scarp degradation over periods of several hundred thousand years.

Offsets by faults may be rapidly erased by erosion or deposition on an active fan surface. Repeated faulting is needed to generate an appreciable offset in face of the competing degrading processes. Fault morphology is thus better preserved on abandoned fan surfaces. The topographic separation on a tectonically disturbed fan surface is always smaller than the total stratigraphic separation which is best traced in excavated trenches. Recurrent faulting events can be recorded in trenches by several colluvial units, by soil horizons, and by several displacements. Measuring the amount of displacement along a ruptured morphology and dating the offset of the markers will constrain the rate of the displacement event and its rate. An active fan surface may, however, be too young to have accumulated many ruptures, whereas ruptures may have been erased on old fan surfaces.

There are different interpretations relating coarsening-upward and gravel progradation to tectonics. One of them is the syntectonic interpretation. A different interpretation (Schlager 1993) postulates a significant time lag between a tectonic event and gravel progradation. According to this view, fan advance extends into periods of tectonic quiescence long after cessation of thrusting (Heller et al. 1988; Blair and Bilodeau 1988; Paola et al. 1992). The greater the distance from the dislocation, the longer the time lag which may extend several tens to a hundred thousand years after the tectonic event.

Influx of coarse clastic sediment is, however by itself, not an evidence of tectonic triggering and can alternatively be attributed to climatic factors which may prove significant even in tectonic controlled areas like the active Himalayan mountain front (Suresh and Kumar 2009). Tectonic quiescence can be concluded based on the absence of uplifted or inclined shorelines, absence of upfan diverging terraces, and lack of offset morphology, of tilted surfaces, or of segmentation of channels' slopes (Hooke 1972).

11.2 Displaced Fans

Fans located in front of mountain fronts, over active strike-slip faults, may be shifted laterally, away from their feeding channel. Different magnitudes of displacement along strike-slip faults indicate different magnitudes or number of earthquake events. Deflected channels or offset of other geomorphic markers on fans can be used to determine the cumulative amount of the horizontal slip. Lithological inheritance from a former sourcing area recorded in the stratigraphy of a fan (Leleu et al. 2009) or an apex that is disconnected from any feeding channel indicate displacement of the fan from its original sourcing area. In order to reconstruct the initial feeding system before the start of any lateral shift and locate each fan in front of its original feeding stream, we need to move the blocks back to their total lateral displacement. This can be determined by a consistent lithologic suitability between the fans and their assumed sourcing drainage basins. The rate of a slip can be calculated from dated sediments. Rates of strike-slip faults range up to a few single mm/year (Nissen et al. 2009). Alluvial fans grow younger along strike-slip faults in the direction of the mountain block shift. Diminishing displacements are expected toward the margins of strike-slip ruptures.

Linear scarps that cross fans are most important geomorphic markers for observing and dating tectonic events and for calculating their slip rate. The age of surface faulting can be based on the youngest faulted alluvial deposits and on the age of the oldest overlying deposits not cut by a scarp-forming fault.

11.3 Subsidence and Stacking

Accommodation space is the space available in a basin for sediment deposition and fan accumulation and can change spatially and volumetrically with the discharge and the flow regime (Paola et al. 1992). Tectonic subsidence creates at the mountain front the basins that serve for accommodation and offers a preservation potential below the incised sourcing basins. With a decrease in the subsidence rate of an accommodation space from moderate, on the order of 1 mm/year, down to 0.01–0.5 mm/year, the rate of sediment delivery from the hinterland may exceed the rate of creation of space, making the accommodation space not enough to store the sediments. Under such conditions, without providing space for deposition, channels will be forced to disperse their load further down, enhancing gravel *progradation* (Viseras et al. 2003; Fig. 9.5b), i.e., growing of the fan forward into the basin where the coarse-grained lobes will spatially alternate with finer local deposits (Clevis et al. 2003) which can be recorded by grain size increases in the stratigraphy (Paola et al. 1992).

Different progradation rates of alluvial fans along a basin axis may mirror different subsidence rates or even reflect a structure of separate sub-basins. Progradation may result from a tectonic pulse, a climatic change, or exposure of different source rock types (Paola et al. 1992). When additional accommodation space becomes available through sinking of the basin and fan supply remains stable, progradation is prevented, and the young, Late Pleistocene or Holocene, alluvial fans are expected to be deposited on top of earlier deposited fans (Fig. 9.5a).

Increased tectonic uplift rate of a source basin located on the *footwall*, i.e., on the uplifted mountain block, will steepen the footwall mountain front, increase the slope gradients of the channels, and thereby strengthen the capacity to detach and transport clasts toward the fan which is located on the *hanging wall*, i.e., on the downfaulted subsiding block which makes up the depocenter (Figs.1.2 and 2.1). Catchments located on footwalls are generally smaller in area, shorter, and steeper and give rise to relatively smaller alluvial fans compared to fans sourced from similar lithologies on hanging walls (Leeder and Jackson 1993).

To allow *stacking* of alluvial fans, i.e., vertical thickening of fans (Fig. 9.5a), a rapid footwall uplift or a hanging-wall basin subsidence should provide much accommodation space adjacent to the mountain front, and a high sediment supply should be drawn into a strong subsiding depocenter that exceeds the aggradation rate. Stacked alluvial fans will be smaller, aggrade with steep depositional slopes, and show less variable facies from the apex down-fan. The stacking fans possess a high width/length ratio and a wide sweep angle. With the waning of mountain uplift, trends change, and incision will exceed the mountain uplift (Bull 2008). Entrenchment that persists through climatic cycles is evidently caused by tectonic forcing. Exposed ancient fans suggest vertical stability and limited subsidence. Under such tectonically inactive conditions, fans tend to have a smaller thickness, smoother slopes, and downward shifting of intersection points and distal deposition.

Young Late Pleistocene or Holocene active fans, burying the older ones, point to subsidence.

Depocenters and the alluvial fans are controlled by the style of faulting. Half-grabens provide an asymmetric accommodation space and display along their margins alluvial fans of different sizes and geometries. Along the steep boundary of a half-graben, which consists of back-tilted blocks, where most drainage is directed away from the rift shoulders, small and steep alluvial fans are formed. The intensive subsidence causes here deep burial of stacking fans (Frostick and Reid 1987; Fig. 11.1.

11.4 Sinuosity of the Mountain Front

The sinuosity index is the ratio of the planimetric length of the mountain front along the mountain-piedmont junction to the length of a straight line joining the ends of the same segment (Bull and McFadden 1977; Keller and Pinter 1996). The mountain-front sinuosity provides an estimation of the degree of balance between stream erosion and weathering that cut embayments into mountains and produces irregular sinuous fronts, relative to the tectonic uplift that tends to generate and keep the original rectilinearity of faulted mountain fronts. Young mountain fronts associated with active uplift are relatively straight, i.e., have a low sinuosity index. When the rate of uplift is reduced or ceases, and the mountain front becomes older, erosion takes over and produces a more embayed front with higher sinuosity values. Decreasing sinuosity values along a mountain front may thus reflect younger ages (Azor et al. 2002). Where mountains rise steeply along range-bounding faults or folds, sinuosity is commonly low. A 1.0–1.5 sinuosity index range indicates low sinuosity. Mountain fronts with sinuosities <2 have been classified as tectonically active fronts (Rockwell et al. 1985; Keller 1986; Silva et al. 2003) showing rates of several m/ky which are considered high. However, low uplift rates of

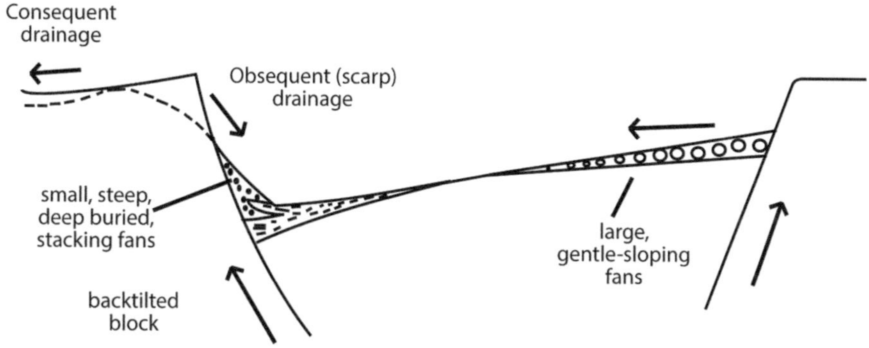

Fig. 11.1 Half-graben provides an example of an asymmetric accommodation space. By the steep boundary-fault small and steep alluvial fans are formed. Subsidence allows for deep burial of stacking fans. On the opposite side is the main drainage input. (Modified after Frostick and Reid 1987)

0.1–0.5 m/ka are sufficient to keep sinuosity index low. Inactive mountain fronts, with a stronger erosive than a tectonic role, will show pronounced front embayments with a high sinuosity index range of 1.9–3.5 and even extreme high index values as 6–7 (Bull and McFadden 1977; Keller and Pinter 1996). Rock resistance and the climate have also their effect but seem not to interfere with the significance of the tectonic-erosional balance shown by the sinuosity index. With time, when the mountain front becomes sinuous, it wears back from its original site where it coincided with the linear bounding-fault trace.

References

Amit R, Zilberman E, Enzel Y, Porat N (2002) Paleoseismic evidence for time dependency of seismic response on a fault system in the southern Arava Valley, Dead Sea rift, Israel. Geol Soc Am Bull 114:192–206

Azor A, Keller EA, Yeats RS (2002) Geomorphic indicators of active fold growth: South Mountain-Oak Ridge anticline, Ventura basin, southern California. Geol Soc Am Bull 114:745–753

Blair TC, Bilodeau WL (1988) Development of tectonic cyclothems in rift, pull-apart, and foreland basins: sedimentary response to episodic tectonism. Geology 16:517–520

Bucknam RC, Anderson RE (1979) Estimation of fault-scarp ages from a scarp-height–slope-angle relationship. Geology 7:11–14

Bull WB (1964) Geomorphology of segmented alluvial fans in western Fresno County, California. USGS No. 352-E

Bull WB (1996) Global climate change and active tectonics: effective tools for teaching and research. Geomorphology 16:217–232

Bull WB (2008) Tectonic geomorphology of mountains: a new approach to paleoseismology, vol 75–116. Blackwell, Malden, pp 117–164

Bull WB, McFadden LD (1977) Tectonic geomorphology north and south of the Garlock fault, California. In: Doehring DC (ed) Geomorphology in arid regions, Proceedings 8th Annual Geomorphology Symposium. State University of New York, Binghamton, pp 115–138

Burbank DW, Anderson RS (2012) Tectonic geomorphology. Wiley- Blackwell, Oxford, pp 35–37, 304-305, 320-327

Clevis Q, de Boer P, Wachter M (2003) Numerical modelling of drainage basin evolution and three-dimensional alluvial fan stratigraphy. Sediment Geol 163(1–2):85–110

Frostick LE, Reid I (1987) Tectonic control of desert sediments in rift basins ancient and modern. In: Frostick LE, Reid I (eds) Desert sediments: ancient and modern, geological society special publication, vol 35. Blackwell, Oxford, pp 53–68

Hanks TC (2000) The age of scarplike landforms from diffusion-equation analysis. In: Quaternary geochronology: methods and applications, pp 313–338

Hanks TC, Wallace RE (1985) Morphological analysis of the Lake Lahontan shoreline and beach-front fault scarps, Pershing County, Nevada. Bull Seismol Soc Am 75:835–846

Harvey AM (1990) Factor influencing quaternary alluvial fan development in southeastern Spain. In: Rachocki A, Church M (eds) Alluvial fans, a field approach. Wiley, Chichester, pp 247–269

Heller PL, Angevine CL, Winslow NS, Paola C (1988) Two-phase stratigraphic model of foreland basin development. Geology 16:501–504

Hooke RL (1972) Geomorphic evidence for late-Wisconsin and holocene tectonic deformation, Death Valley, California. Geol Soc Am Bull 83:2073–2098

Keller EA (1986) Investigation of active tectonics: use of surficial earth processes. In: Active tectonics, pp 136–147

Keller EA, Pinter N (1996) Active tectonics. Prentice Hall, Upper Saddle River, pp 138–147

Leeder MR, Jackson JA (1993) The interaction between normal faulting and drainage in active extensional basins, with examples from the western United States and Central Greece. Basin Res 5:79–102

Leleu S, Ghienne JF, Manatschal G (2009) Alluvial fan development and morpho-tectonic evolution in response to contractional fault reactivation (Late Cretaceous – Palaeocene), Provence, France. Basin Res 2:157–187

Nichols GJ (1987) Syntectonic alluvial fan sedimentation, southern Pyrenees. Geol Mag 124:121–133

Nissen E et al (2009) The late Quaternary slip-rate of the Har-Us-Nuur fault (Mongolian Altai) from cosmogenic 10 Be and luminescence dating. Earth Planet Sci Lett 286(3):467–478

Paola C, Heller PL, Angevine C (1992) The large-scale dynamics of grain size variations in alluvial basins I. Theory. Basin Res 4:73–90

Pinter N, Keller EA (1995) Late quaternary deformation in the northern Owens Valley, California: geomorphic analysis of tectonic tilt. Geol Rundsch 84:200–212

Rockwell TK, Keller EA, Johnson DL (1985) Tectonic geomorphology of alluvial fans and mountain fronts near Ventura, California. In: Tectonic geomorphology. In: Proceedings of the 15th annual geomorphology symposium. Allen and Unwin Publishers, Boston, pp 183–207

Schlager W (1993) Accommodation and supply – a dual control on stratigraphic sequences. Sediment Geol 86:111–136

Silva PG, Goy JL, Zazo C, Bardajı T (2003) Fault-generated mountain fronts in southeast Spain: geomorphologic assessment of tectonic and seismic activity. Geomorphology 50(1):203–225

Suresh N, Kumar R (2009) Variable period of aggradation and termination history of two distinct Late Quaternary alluvial fans in the Soan Dun NW Sub-Himalaya: impact of tectonics and climate. Himal Geol 30:155–165

Viseras C, Calvache ML, Soria JM, Fernandez J (2003) Differential features of alluvial fans controlled by tectonic or eustatic accommodation space, examples from the Betic Cordillera, Spain. Geomorphology 50:181–202

The Climatic Control 12

Abstract

Alluvial fans are not restricted to any specific climate. Climatic variations within tectonic cycles may produce sufficient fluctuations in water and sediment discharge to change trends from aggradation to entrenchment. The sedimentary and the morphologic signatures following climatic events can be similar to those created by tectonism. Regional synchroneity of the fan behavior may be the signal implying climatic forcing.

Climate controls the vegetative cover and the water/sediment yield ratio that determine whether fans undergo aggradation or degradation. Vegetation in the sourcing area maintains the slope stability by favoring infiltration and inhibiting runoff thus decreasing the supply of sediment delivered to the fan. Dry climates, deforestation, overgrazing, and soil tillage diminish the soil moisture and the vegetation density, increasing slope erosion in the sourcing area and enhancing fan growth. The humid model links fan aggradation to wetter conditions with greater runoff generation and slope instability leading to an intensive supply of sediments.

Keywords

Vegetation cover · Water discharge/sediment supply ratio · Effective precipitation · Paraglacial fans · Humid/arid impacts

Tectonism is considered to impose a primary control on formation of fans by providing the basic conditions required for fan deposition, i.e., relief and accommodation space. Within tectonic cycles, climatic changes may exert sufficient fluctuations in water and sediment discharge to change trends from aggradation to entrenchment (Harvey and Wells 1994; McDonald et al. 2003). Even in tectonically active

© Springer Nature B.V. 2019

D. Bowman, *Principles of Alluvial Fan Morphology*,
https://doi.org/10.1007/978-94-024-1558-2_12

mountain fronts, changes to humidity or aridity may become an important secondary control of alluvial fans. Climate controls alluvial fans through weathering intensity and the sediment production rate in the sourcing area as well as through the influence on the vegetation cover, on the runoff, and on the sediment transport. Climate change is a driving force in fan development because the lifetime of alluvial fans exists long enough to encounter long-term climatic fluctuations (Ritter et al. 1995). Short-term, decadal- to centennial-scale climatic fluctuations, as well as episodic events, may superimpose their mark on the effects of the slow, long-term climatic changes. It should be noted, however, that climatic events can produce sedimentary and morphologic signatures similar to those created by tectonism, as, for example, abrupt changes in sediment mobilization. However, changes in processes, as from debris flows to dilute flows, need not reflect climatic changes but may as well result from intrinsic changes in the sourcing catchment. To prove a climatic control requires that the changes be spatially widespread and synchronous, irrespective of lithology or of the relief of the sourcing drainage basin. However, a tectonic event may show as well across a wide area of adjacent fans. Widely spread tectonic activity along mountain fronts makes it sometimes difficult or even impossible to be distinguished from the effect of climate.

Alluvial fans are not restricted to any specific climate but are spread throughout widely variable climatic environments: arid, semiarid, humid-temperate, tropical, alpine, and proglacial climatic zones. Fans may thus be densely or partly forested or completely bare.

12.1 The Effect of Vegetation and of the Water/Sediment Ratio

Precipitation may change by the total annual volume as well as by the storm frequency or intensity, which might increase even in the case of a decrease in the total volume. Increase in sediment yield may often be largely due to increase in *extreme storm events*. *Effective precipitation* is the amount of water, after losses to infiltration, transpiration, etc., that remains available to support plants, weather rocks, form soils, supply runoff, and determine the sediment yield from the catchment area and the sediment/water ratio (Bull 1991; Mack and Leeder 1999). Small availability of sediment leads to increased stream power. When sediment load decreases and water discharge or the gradient increases, stream power becomes more than sufficient for transporting the load, and degradation will start. Under high sediment supply conditions, when stream power becomes insufficient, the water is unable to take up all material, and aggradation occurs.

The intensity of the vegetation cover in sourcing drainage basins influences the sediment yield and has a strong impact on the evolution of fans, in both aggradational and incisional regimes (Bull 1991). Vegetation is most dense and diverse on active fans and on low terraces watered by overbank flow. Vegetation is low in streambeds and limited on old fan surfaces that receive only direct rain. A mature fan surface which is covered by a well-developed desert pavement has no vegetation cover or only a very sparse one, although in the channels it is possible to find some perennials. In the sourcing drainage basins, vegetation maintains the slope stability by favoring infiltration, decreasing runoff, retarding flow, and thus decreasing the supply of water and of sediments. Enhancing vegetation decreases slope stripping in the sourcing basins and increases the buildup of deeply weathered slopes.

The root web of trees on fans provides reinforcement against erosion and channel incision. Exposed roots indicate the erosional trends and magnitudes. Scars on trunks result from abrasion. Boulders and gravel may be wedged between the trunks of the trees, indicating the level of the floods. Trees obstruct and store debris and logs on the upslope side and may create a stepped morphology up to a few meters high. The trees on forested fans may as well be buried by sediments and killed (Wilford et al. 2005; Giles et al. 2010).

Active washes favor botanical communities dominated by annuals. Plants on active fans trap, during sheet flows, sand and fines which provide a favorable environment for microbiotic soil crust development. Trapping moisture and fixing nitrogen assist the growth of additional annuals and perennials. Drier climates cause a reduction in soil moisture, loss of the vegetation cover, and depletion of the materials stored in the slopes of the sourcing basins (Bull 1991, 2000). The sediments that mantled hillslopes during wetter regimes will be transported to the fans during transitions to drier periods. Fan aggradation is facilitated by deforestation, by overgrazing, and by soil tillage, all of which destabilize the sediment cover on slopes of the sourcing catchments and release an increased volume of sediments toward the fans (Bull 1991; Leader et al. 1998).

The anthropogenic hypothesis recognizes that vegetation modification and agricultural activity lower the threshold for soil erosion, leading to gullying during periods of high runoff and to fan aggradation (Anderson et al. 2000). The coincidence in the timing of fan aggradation and climate changes along with the timing of land use has still to be examined (Brazier and Ballantyne 1989).

12.2 The Paraglacial Impact

Glacial, paraglacial, and interglacial alterations had a fundamental impact on the evolution of alluvial fans (Weissmann et al. 2002) and exemplify the impact of the ratio of sediment load to water discharge. Through active freeze-thaw weathering and by erosion of the bedrock, glaciation produced large amounts of detrital material. During the glacial cold and dry period, the ratio of sediment load to water discharge was high, and the glaciated drainage basins supplied an increased sediment discharge (Ritter et al. 1993). Following the Late Pleistocene deglaciation, the valley floors became ice-free, though still subject to a *paraglacial* regime. The term *paraglacial* incorporates non-glacial processes that started upon deglaciation and were still conditioned by the former glaciation. These processes extended well into the non-glacial interval and operated over timescales of about 10^4 years. In this phase of retreating glaciers, the unconsolidated glaciogenic sediment stores, i.e., morainic debris, tills, and glaciolacustrine deposits, were exposed to erosion and provided the major reserves of unconsolidated sediments to be transported for fan deposition (Ashmore 1993; Ballantyne 2002; Church and Ryder 1972, 1989; Ryder 1971).

The withdrawal of the glaciers enhanced the sediment release at rates greatly exceeding the denudation rates. The drainage basins with former glacial erosion and mechanical production of sediments contributed more debris than would be expected from their dimension. The fans that were derived from these abundant unconsolidated sediments are known as *paraglacial fans*. Maximum sediment sourcing occurred very soon after exposure of the land surface from under the retreating glacier ice with the rates dropping rapidly during deglaciation (Church and Ryder 1972). The sources of glaciogenic sediments became gradually exhausted, resulting in diminishing sediment delivery. The declining availability of sediment caused a change in the sediment/water ratio, and more and more paraglacial fans have ceased to grow and are now entrenched.

12.3 Humid and Arid Impacts

Humid Conditions cause more weathering, more production of colluvium, and more soil development. High-intensity precipitation increases sediment supply through gullies and major slope failures in the sourcing basin. A high water discharge to sediment supply ratio will result in an increased stream power, increasing thereby the transporting effectiveness in the drainage basin and the supply of sediments to fans. This *humid model* interprets fan aggradation to coincide with wetter conditions, more runoff generation, and slope instability, which can all be associated with glacial periods (Harvey et al. 1999a, b; Ponti 1985). A different model of fan formation, with climate transition to wetter conditions, can be found in deserts when changing from superarid to arid. Wetter conditions increase the effective precipitation in conditions of sparse vegetation, denude the catchment of debris, and increase the sediment yield. Such transition to a wetter climate may, however, trigger, gradually, a complex response due to a gradual increase in the vegetation cover that will counteract hillslope depletion. When the stream power keeps high on the fans (Bull 1979) as during the pluvial periods, fan incision will dominate.

Arid Impacts When less moisture is available to support the protective hillslope vegetation, more sediments are moved from the sourcing basin. Brief, high-intensity flood events, typical of arid environments, strip more colluvium from the sparsely vegetated hillslopes of the sourcing basins to be delivered to the fans. Droughts that increase the frequency of forest fires trigger an increased sediment production and enhance accumulation of fans (Eppes and McFadden 2008). Spelz et al. (2008) suggested that major phases of fan aggradation coincided with Quaternary cold phases that were too dry to support woodland vegetation but humid enough to strip hill slopes. The aggradation trends on fans are likely to persist until the deposition-erosion threshold is crossed, i.e., when sediment concentration in flows decreases, either due to depletion of the colluvium on the hillslopes (Bull 1979, 2000) or when a denser vegetation cover, following more humid conditions, causes the sediment to remain stored.

Fan studies suggest a temporal link between trends of aggradation or entrenchment and past climates. Periods of fan aggradation have been related to relatively more humid periods that coincided with increased storminess such as the Little Ice Age of the sixteenth to the nineteenth centuries throughout Central Europe (Rumsby and Macklin 1996). Stages of fan entrenchment have been related to lengthy arid periods *but also vice versa*, i.e., to less sediment supply during interglacials, because of an increased vegetation cover. The relation of aridity and humidity to aggradation and entrenchment seems still unclear.

Distinct wetter or drier periods have been suggested to be recorded also by changes in the *stratigraphic record* (Waters et al. 2010). Deep- and large-scale conglomerate-filled-channels with strong stream-regime-characteristics, including imbrication, open framework, crudely horizontally stratification and clast-supported gravels, have been interpreted as indicating increased rainfall and strong floods with perennial channels. Drier periods have been marked by sandier braided fluvial deposits and pedogenic calcretes.

References

Anderson E, Harrison S, Passmore DG, Mighall TM (2000) Holocene alluvial-fan development in the Macgillycuddy's Reeks, Southwest Ireland. Geol Soc Am Bull 112(12):1834–1849

Ashmore P (1993) Contemporary erosion of the Canadian landscape. Prog Phys Geogr 17:190–204

Ballantyne CK (2002) Paraglacial geomorphology. Quat Sci Rev 21:1935–2017

Brazier V, Ballantyne CK (1989) Late Holocene debris cone evolution in Glen Feshie, western Cairngorm Mountains, Scotland. Earth and Environ Sci Trans R Soc Edinb 80(1):17–24

Bull WB (1979) Threshold of critical power in streams. Geol Soc Am Bull 90(5):453–464

Bull WB (1991) Geomorphic responses to climatic change. Oxford University Press, Oxford

Bull WB (2000) Correlation of fluvial aggradation events to times of global climate change. In: Quaternary geochronology: methods and applications. American Geophysical Union, Washington, DC, pp 456–464

Church M, Ryder JM (1972) Paraglacial sedimentation: a consideration of fluvial processes conditioned by glaciation. Geol Soc Am Bull 83:3059–3071

Church M, Ryder JM (1989) Sedimentology and clast fabric of subaerial debris flow facies in a glacially influenced alluvial fan – a discussion. Sediment Geol 65:195–196

Eppes MC, McFadden L (2008) The influence of bedrock weathering on the response of drainage basins and associated alluvial fans to Holocene climates, San Bernardino Mountains, California USA. Holocene 18(6):895–905

Giles P, Nichols G, Wilford D (2010) Alluvial fans: from reconstructing past environments to identifying contemporary hazards. Geomorphology 118(1–2):224

Harvey AM, Wells SG (1994) Late Pleistocene and Holocene changes in hillslope sediment supply to alluvial fan systems: Zzyzx, California. In: Millington C, Pye K (eds) Environmental change in drylands. Wiley, Chichester

Harvey AM, Wigand PE, Wells SG (1999a) Response of alluvial fan systems to the late Pleistocene to Holocene climatic transition: contrasts between the margins of pluvial Lakes Lahontan and Mojave, Nevada and California, USA. Catena 36(4):255–281

Harvey AM, Silva PG, Mather AE, Goy JL, Stokes M, Zazo C (1999b) The impact of Quaternary sea-level and climatic change on coastal alluvial fans in the Cabo de Gata ranges southeast Spain. Geomorphology 28(1):1–22

Leeder MR, Harris T, Kirkby MJ (1998) Sediment supply and climatic change: implication for basin stratigraphy. Basin Res 10:7–18

Mack GH, Leeder MR (1999) Climatic and tectonic controls on alluvial-fan and axial fluvial sedimentation in the Plio-Pleistocene Palomas half graben southern Rio Grande rift. J Sediment Res 69:635–652

McDonald EV, McFadden LD, Wells SG (2003) Regional response of alluvial fans to the Pleistocene-Holocene climatic transition, Mojave Desert, California In: Enzel Y, Wells SG, Lancaster N (eds) Paleoenvironments and paleohydrology of the Mojave and Southern Great Basin Deserts. Geological Society of America Special Paper, vol 368, pp 189–205

Ponti DJ (1985) The quaternary alluvial sequence of the Antelope Valley, California. In: Geological Society of America Special Papers, vol 203, pp 79–96

Ritter JB et al (1993) Quaternary evolution of Cedar Creek alluvial fan, Montana. Geomorphology 8:287–304

Ritter JB, Miller JR, Enzel Y, Wells SG (1995) Reconciling the roles of tectonism and climate in Quaternary alluvial fan evolution. Geology 23:245–248

Rumsby BT, Macklin MG (1996) River response to the last neoglacial (the 'Little Ice Age') in northern, western and central Europe. Geological Society, London, Special Publications 115(1), 217–233

Ryder JM (1971) Some aspects of the morphometry of paraglacial alluvial fans in south-central British Columbia. Can J Earth Sci 8:279–298

Spelz RM, Fletcher JM, Owen LA, Caffee MW (2008) Quaternary alluvial-fan development, climate and morphologic dating of fault scarps in Laguna Salada, Baja California, Mexico. Geomorphology 102:578–594

Waters JV, Jones SJ, Armstrong HA (2010) Climatic controls on late Pleistocene alluvial fans, Cyprus. Geomorphology 115(3):228–251

Weissmann GS, Mount JF, Fogg GE (2002) Glacially driven cycles in accumulation space and sequence stratigraphy of a stream-dominated alluvial fan, San Joaquin Valley, California, USA. J Sediment Res 72:240–251

Wilford DJ, Sakals ME, Innes JL, Sidle RC (2005) Fans with forests: contemporary hydrogeomorphic processes on fans with forests in west central British Columbia, Canada. In: Harvey AM, Mather AE, Stokes M (eds) Alluvial fans. Geomorphology, sedimentology and dynamics, Special Publication, vol 251. The Geological Society, London, pp 25–40

The Base-Level Effect 13

Abstract
Base-level is an imaginary horizontal surface to which the fluvial system strives to erode. A distinction is made between a terminal and a local base-level. The base-level of many alluvial fans is an inland, local, intermountain basin. Its margins, which are the typical site of fan deposition, are not base-level loci. A falling base-level will trigger entrenchment only if the new exposed area will be steeper than the threshold gradient for erosion. Intersection points that are the end spot of an entrenched fan segment do not function as a base-level. Profile shortening by toe trimming causes headward entrenchment although not a base-level drop.

Keywords
Terminal/local base-level · Knickpoint · Intersection point · Toe trimming

Base-level is an imaginary horizontal surface which marks the lowest limit of fluvial activity. The base-level is the sink of drainage systems, i.e., the lowest elevation where channels deposit and to which the fluvial systems strive to erode. At the base-level, the hydraulic gradient is reduced to zero, and fluvial processes cease (Leopold and Bull 1979). A distinction is made between the *terminal* or *ultimate base-level*, which is the stable, global sea level and *local endorheic base-levels* such as playas, lakes, or inner continental basins.

An axial master stream along a basin typically serves as a local base-level, drawing toward it the alluvial fan runoff. A base-level in form of a basin bottom may rise or sink due to local intra-basin faulting or folding, and a base-level in the form of a lake level may rise or sink as a result of climatic/hydrological changes. Resistant outcrops which function as lithologic bedrock barriers and often operate as local

© Springer Nature B.V. 2019
D. Bowman, *Principles of Alluvial Fan Morphology*,
https://doi.org/10.1007/978-94-024-1558-2_13

base-levels are usually not relevant to alluvial fans which are typically composed of more erodible deposits.

In all cases the base-level represents an *external forcing* on the fan system. The different base-level controls over alluvial fans are exemplified by the following scenarios:

1. Alluvial fans typically grow along piedmonts on the margin of basins which, however, are not a base-level site. When all the sediments are deposited on the fan and not transported beyond its boundaries and most of the water infiltrates, the fan as a sedimentary body stays completely *above the base-level* and is decoupled from it. Its evolution is in no way controlled by the base-level. It is dominated by its sourcing drainage basin, i.e., *driven top-down* (Mather et al. 2017).
2. Prograding fans often fail to extend and keep contact with the rapidly retreating lake at their front. The toe of the fan does not touch the receding lake that consequently exerts no control on the fan evolution except by providing a wider accommodation space (Bowman 1988).
3. When fans are deeply dissected all the way from the apex down to the toe and beyond, base-level lowering will trigger *bottom-up driven entrenchment*. Low sediment input may enhance the erosive effect of such lowering base-level. The narrower the flow, the greater will be the impact of the base-level lowering. If low cohesiveness of the fan alluvium promotes channel widening, the lower stream power will decrease the erosional base-level effect (Schumm 1993, 2005).
4. The gradient of the exposed area following a lake level fall will control the base-level effect (Harvey 2005; Holbrook and Schumm 1999). Incision will start only if the gradient exposed by the base-level fall is steeper than the threshold gradient for erosion; otherwise flow may simply prograde on the newly exposed frontal area without entrenchment, controlled by thesediment flux.
5. The impact of a falling base-level dies out when erosion propagates upfan as a time transgressive effect. Where a rejuvenated entrenched reach merges with the upstream fan surface, a *knickpoint* will be formed (Harvey 2004, 2011). Such knickpoint does not constitute a base-level and may be mistaken for being generated tectonically.
6. Intersection points along channels (Fig. 8.2) are knickpoints that designate the lowest entrenched point along a segment and do not constitute a base-level (Leeder 1978). Intersection points are not controlled by *bottom-up headward erosion* dictated by a lowering base-level.
7. Lateral swing of an axial, non-incising trunk river toward the distal margins of a fan will cause distal *toe trimming* of the fan, forming a cliff and thereby causing profile shortening of the fan slope and establishing a bottom-up, headward entrenchment process (Leeder and Mack 2001) activated *without any base-level lowering*. Gullies and channels extend up from the toe-cut cliff line, become gradually shallow, and die out or cause intensive fan erosion.
8. A base-level rise in the form of a rising lake level will trigger upward migration of lacustrine deposition forming a stratigraphic onlap veneer (Schumm 1993).

The extent of base-level dominance is strongly related to the evolutional phase of the alluvial fan: braiding and strong infiltration with minor flow beyond the fan boundaries are conditions that characterize the depositional phase of fan evolution characterized by a weak base-level control. The effect of base-level is strongest when a concentrated channelized flow crosses the fan, typical to an advanced incisional stage.

References

Bowman D (1988) The declining but non-rejuvenating base-level, the Lisan Lake, the Dead Sea area, Israel. Earth Surf Process Landf 13:239–249

Harvey AM (2004) The response of dry-region alluvial fans to late quaternary climatic change. Desertification in the third millenium. Balkema, Rotterdam, pp 83–98

Harvey AM (2005) Differential effects of base-level, tectonic setting and climatic change on quaternary alluvial fans in the northern Great Basin, Nevada, USA. Geol Soc Lond, Spec Publ 251(1):117–131

Harvey A (2011) Dryland alluvial fans. In: Thomas DSG (ed) Arid zone geomorphology, process, form and change in drylands, 3rd edn. Wiley, Chichester, pp 333–371

Holbrook J, Schumm SA (1999) Geomorphic and sedimentary response of rivers to tectonic deformation: a brief review and critique of a tool for recognizing subtle epeirogenic deformation in modern and ancient settings. Tectonophysics 305(1):287–306

Leeder MR (1978) A quantitative stratigraphic model for alluvium, with special reference to channel deposit density and interconnectedness. In: Miall AD (ed) Fluvial sedimentology. Memoir, Canadian Society of Petroleum Geologist, pp 587–596

Leeder MR, Mack GH (2001) Lateral erosion (toe cutting) of alluvial fans by axial rivers: implications for basin analysis and architecture. J Geol Soc Lond 158:885–893

Leopold LB, Bull WB (1979) Base level, aggradation, and grade. Proc Am Philos Soc 123(3):168–202

Mather AE, Stokes M, Whitfield E (2017) River terraces and alluvial fans: the case for an integrated quaternary fluvial archive. Quat Sci Rev 166:74–90

Schumm SA (1993) River response to base level implications for sequence stratigraphy. J Geol 101:279–294

Schumm SA (2005) River variability and complexity. Cambridge University Press, Cambridge

Groundwater

14

Abstract

Alluvial fans in piedmont areas are major reservoirs of groundwater. Fan aquifers are typically complex systems composed of multiple aquifers and aquitards. The highest infiltration rates on alluvial fans are at the apex zone. Farther downfan where the beds become thinner and sorting improves, there is a decrease in the hydraulic conductivity. The aquitards act as partial barriers to infiltration, confine vertical percolation, and direct the groundwater flow laterally into separate strata, thus building individual pressurized horizons. The aquifer transmissivities and the storage coefficients are highest in the fans' intermediate zone.

As the groundwater carries part of the granular load, withdrawal of groundwater from the fan aquifer causes compaction of the underground grain material and reduces the pore volume. The removal of groundwater from the fan aquifer results in consolidation of the deposits and causes ground subsidence. Changes of groundwater level have also been observed to accompany earthquakes.

Keywords

Compartmentalization · Complex aquifer-aquitard system · Recharge · Contamination · Pore fluid pressure · Subsidence

Alluvial fans comprise a coarse hydrological province that is an important groundwater reservoir in piedmont zones. The fans consist of sequences of gravel, sand, mud, and paleosols and are typically heterogeneous in their hydraulic conductivity, i.e., in the ability to transmit water through pore spaces. The permeable gravelly composition of fans endows them with high transmissivity and pumpage capacities. Their vadose zone permits significant recharge, both natural and artificial.

© Springer Nature B.V. 2019
D. Bowman, *Principles of Alluvial Fan Morphology*,
https://doi.org/10.1007/978-94-024-1558-2_14

Being directional landforms, the alluvial fans show a decrease in the hydraulic conductivity down-fan where the beds thin. The hydraulic conductivity may vary down-fan over ten orders of magnitude (Neton et al. 1994). When rich with debris flows, transmissivity – the rate at which groundwater is transmitted horizontally – is low in the proximal fan area and increases to highest values in the mid-fan. When devoid of debris flows, the highest infiltration rates on an alluvial fan are to be found at the apex, where the coarsest sediment, although poorly sorted, is deposited. Infiltration is especially high in incised stream channels where from fines have been winnowed. From the proximal area down-fan, the permeability at the fan surface decreases, and the water table shallows. The mid-fan aquifer has the highest transmissivities and storage coefficients. Its deposits consist of highly stratified, well-sorted beds that show the greatest lateral continuity (Cehrs 1979). The distal parts of fans are enriched by fine-grained components with lower transmissivity and storage coefficients compared with the mid-fan zone. A certain impediment to the conductivity of fans may be the precipitation of intergranular carbonate, silica, or iron oxides. Alluvial fans are generally an oxidizing environment and are often impregnated with Fe oxides.

14.1 Compartmentalization

The fan aquifers often separate from the head into a complex radiating pattern of subterranean paths (Hung et al. 2010), making a *complex aquifer-aquitard system* composed of a multiple aquifer structure. The upper aquifer is mostly unconfined, while the several lower ones are confined, separated by aquitards of clay to fine sand of low permeability (Fig. 14.1). The aquitards are originating from interbedded lacustrine or marine deposits, from overbank sheetwash, or from extensive paleosol horizons. Debris-flow deposits too may form leaky confining units (Cehrs 1979). The horizons of low conductivity and significant perching radiate into separate subterranean branches and are the cause of the notable compartmentalization of the fan aquifer system (Munevar and Marino 1999; Weissmann et al. 2013). The aquitards which contribute to the compartmentalization of the fan aquifer are often less effective and incomplete as barriers and allow reconnections through the gravel, through the sand lenses, and through channel fills.

The aquitards, which develop mainly in the mid- and distal fan areas, are laterally more or less continuous and act as partial barriers to infiltration. The aquitards are partway confining the vertical percolation and directing water laterally through the aquifers, thus building individual pressurized horizons. Within the inclined stratigraphy of fans, such hydraulic structure allows artesian conditions. The heterogeneity of the groundwater in alluvial fans shows through the wide mix of water with various residence times, broad age distributions, and spatially variable groundwater velocities. When aquitards are noncontinuous but leave good interconnections between the different aquifer levels, the fan aquifer behaves more as a homogeneous

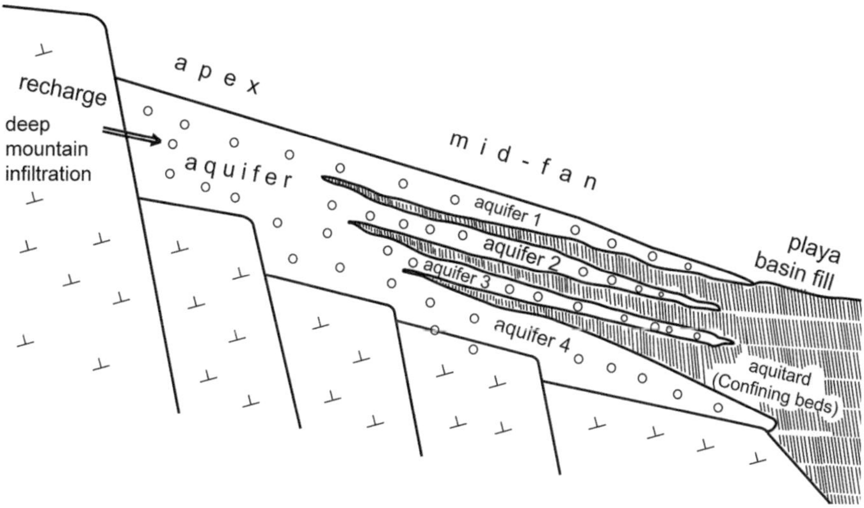

Fig. 14.1 Distribution of aquifers and aquitards in an alluvial fan, schematic. (Modified after Liu et al. 2002). The effect of debris-flow deposits is not shown

unconfined system with greater hydrochemical homogeneity and more even distribution of isotopic and geochemical parameters (Stimson et al. 2001).

The directions of flow in an alluvial aquifer follow the permeability of the alluvium and the gradient of the subsurface layers. Connectivity of aquifer bodies within the subsurface is gained through aquifer corridors that consist of fluvial channel stacked deposits, sometimes of multistory sand and gravel bodies that are surrounded by non-aquifer material. Greater degree of connectivity between potential aquifer bodies in the across-fan direction, normal to the transport directions, is expected near the fan apice where the aquifer bodies are to converge. Thin alluvial fans on relatively impermeable bedrock push the water table upward in close proximity to the fan surface. The arrest of downward percolation increases the lateral flow throughout the aquiferous layers. At the distal parts of fans, the presence of brines under playas often forces the fresh water to exit above and form a spring line or broad wetlands which support rich vegetation. The high water table and the low gradient of the fan's margin form waterlogged and marshy terrains with organic-rich anoxic deposits.

Reconstruction of the subsurface stratigraphy of alluvial fans by modeling should be based on geomorphic knowledge of fluvial channel geometry, avulsion processes, depositional patterns, and likely dimensions of aquifer bodies and should fit data from multiple well locations in the hosting fan. Model simulations are expected to yield potential channel pathways within a discrete depth range and probability maps for finding aquifer locations within a series of depth slices. Stacking the depth slices together gives information on the likely spatial distribution of aquifers in the subsurface (Van Dijk et al. 2016).

14.2 Contamination, Recharge, and Tectonic Effects

Water supplied from the sourcing drainage area infiltrates and recharges the uppermost unconfined fan aquifer. Additional sources of the groundwater in unconfined fan aquifers are infiltration from direct precipitation, irrigation return flow, and groundwater inflow from bounding mountain slopes. The groundwater level starts to rise in direct response to floods and will continue for months owing to the time lag of vertical percolation. Hydraulic calculations in the Atacama Desert confirm that groundwater may be transmitted from the fan apex to a radial distance of about 40 km within 20–30 years (Houston 2002). The mean residence times of groundwater in fans may extend to an order of several tens of years.

The highly permeable fan aquifers are very vulnerable to anthropogenic contamination, such as chemicals, pesticides, or fertilizers carried in surface runoff from farmlands, landfills, sewage systems, and industrial effluents. Soluble contaminants may remain and accumulate in the less permeable, silty to clayey sediments. Exposure of alluvial aquifers to contaminants may lead to gradual rising concentrations that become irreversible on the scale of decades to centuries. In alluvial fans that debouch into the sea, seawater may intrude into the aquifer and cause groundwater salinization, which will advance upfan if the aquifer is overexploited by pumpage. Artificial recharge of alluvial fan aquifers may utilize runoff that is otherwise wasted, by injection wells or by surface spreading accomplished by constructing dams, by infiltration ponds, and by trenches from which the water will infiltrate into the highly permeable alluvium and extend the time for infiltration. To decrease sediment deposition in the spreading areas, sediment traps can be built upstream. Artificial recharge may show some impact on seawater intrusions and decrease high chloride in wells.

Pore fluid pressure can be very sensitive to crustal strains induced by tectonics. Oscillatory and steplike changes of groundwater level were found to accompany earthquakes. The area where groundwater level rose or fell coseismically is generally consistent with the area of volumetric compression or expansion. Fractures may open and lead to increased permeability and account for a decrease in the water level. Changes in groundwater levels in alluvial fans induced by the ML 7.3 Chi-Chi earthquake of 1999 ranged from 1.0 to 11.1 m. Coseismic changes of groundwater level were recorded in confined, partially confined, and unconfined aquifers. The recovery of coseismic water-level changes usually took from less than 1 h up to months in confined aquifers. The amplitude of coseismic water-level rises could be related to the magnitude of the earthquake and the well-epicenter distance (Chia et al. 2001; Roeloffs 1998).

14.3 Subsidence

Groundwater carries a part of the granular load. When the support provided by the fluid pressure is reduced, following withdrawal of groundwater, the stress of the weight of the overlying rock is transferred to the granular system. Because the granular structure of the aquifer system is not rigid, rearrangement results in reduction of the pore volume because of compaction. Most of the compaction occurs in fine-grained clay aquitard deposits, with their substantially greater porosity and compressibility though lesser permeability compared to the more transmissive, coarser-grained sand and gravel layers (Galloway et al. 1999; Hung et al. 2010). The areas of maximum subsidence must not necessarily coincide with areas of maximum water-level decline but may be related to the larger thickness of the fine-grained, compressible sediments. Differential land subsidence creates tensional stresses that may ultimately result in fissuring. Ground subsidence and layer compression in fans are also affected by rainy and dry seasons.

Land subsidence accompanied with heavy withdrawal of the groundwater, and not only on fans, has become a worldwide problem (Liu et al. 2004). Land subsiding rates of 10 cm/year have been reported by Chen et al. (2010). Withdrawing groundwater from alluvial fan aquifers may cause total subsidence of the fan surface up to a few meters.

References

Cehrs D (1979) Depositional control of aquifer characteristics in alluvial fans, Fresno County, California: summary. Geol Soc Am Bull 90:709–712 1282–1309

Chen CH, Wang CH, Hsu YJ, Yu SB, Kuo LC (2010) Correlation between groundwater level and altitude variations in land subsidence area of the Choshuichi alluvial fan, Taiwan. Eng Geol 115(1):122–131

Chia Y, Wang YS, Chiu JJ, Liu CW (2001) Changes of groundwater level due to the 1999 Chi-Chi earthquake in the Choshui River Alluvial Fan in Taiwan. Bull Seismol Soc Am 91(5):1062–1068

Galloway DL, Jones DR, Ingebritsen SE (1999) Land subsidence in the United States. U.S. Geological Survey 1182, 177p

Houston J (2002) Groundwater recharge through an alluvial fan in the Atacama desert, Northern Chile: mechanisms, magnitudes and causes. Hydrol Process 16:3016–3035

Hung WC, Hwang C, Chang CP, Yen JY, Liu CH, Yang WH (2010) Monitoring severe aquifer-system compaction and land subsidence in Taiwan using multiple sensors: Yunlin, the southern Choushui River Alluvial Fan. Environ Earth Sci 59:1535–1548

Liu CW, Jang CS, Chen SC (2002) Three-dimensional spatial variability of hydraulic conductivity in the Choushui River alluvial fan, Taiwan. Environ Geol 43:48–56

Liu CH, Pan YW, Liao JJ, Huang CT, Ouyang S (2004) Characterization of land subsidence in the Choshui River alluvial fan, Taiwan. Environ Geol 45(8):1154–1166

Munevar A, Marino MA (1999) Modeling analysis of groundwater recharge potential on alluvial fans using limited data. Groundwater 37(5):649–659

Neton MJ, Dorch J, Olson CD, Young SC (1994) Architecture and directional scales of heterogeneity in alluvial fan aquifers. J Sed Res B64:245–257

Roeloffs EA (1998) Persistent water level changes in a well near park- field, California, due to local and distant earthquake. J Geophys Res 103:869–889

Stimson J, Frape S, Drimmie R, Rudolph D (2001) Isotopic and geochemical evidence of regional scale anisotropy and interconnectivity of an alluvial fan system, Cochabamba Valley, Bolivia. Appl Geochem 16:1097–1114

Van Dijk WM, Densmore AL, Sinha R, Singh A, Voller VR (2016) Reduced-complexity probabilistic reconstruction of alluvial aquifer stratigraphy, and application to sedimentary fans in northwestern India. J Hydrol 541:1241–1257

Weissmann GS, Hartley AJ, Scuderi LA, Nichols GJ, Davidson SK, Owen A, Nordt LC (2013) Prograding distributive fluvial systems: geomorphic models and ancient examples. New Front Paleopedol Terr Paleoclimatol: SEPM Spec Publ 104:131–147

15 Morphology of the Fan Surface

Abstract

Vertical incision and lateral shifting of channels across the fan form a characteristic stepped fan morphology. The incised channels are often punctuated by backfillings that become with time entrenched again and finally display inset terraces. The fan surface displays both grain roughness and a microtopography roughness which gradually diminish with time by surface smoothing processes. The roughness can be shown by remote sensing through the different brightness of the fan surface. Estimation of the surface roughness can be accomplished by conducting spectral field measurements, by airborne hyperspectral data analysis, by measurements of the ground-reflected solar radiance, and by using airborne lidar-radar data. The trend of decreasing roughness with age suggests the roughness criteria as a measure of relative-age determination. The average backscatter from a surface can be effectively utilized as a proxy for the surface age.

Keywords

Channel backfilling · Inset fan terrace · Strath terrace · Stepped fan morphology · Telescopic pattern · Grain roughness · Microtopography · Surface smoothing

The fan surface tells much about its evolution and about the recent trends. The trends of aggradation and dissection are also well reflected in the stratigraphy. The entrenched, proximal, oldest fan surface has direct precipitation the only source of runoff. The distal down-fan and toe zones are transitional between the alluvial fan and the axial river channel belt environment. The fan toe is generally the site of potential intercalations of axial-fluvial and alluvial fan deposits. The toe may also show as a sharp scarp termination reflecting a lateral toe cutting. The alluvial fan may as well prograde across a toe scarp (Leeder and Mack (2001). Lack of

© Springer Nature B.V. 2019

D. Bowman, *Principles of Alluvial Fan Morphology*,
https://doi.org/10.1007/978-94-024-1558-2_15

intercalation of alluvial fans and fluvial axial sourced material may reflect buffering within the fan.

The alluvial fans are often typified by poorly sorted, coarser, angular to subangular deposits dominated by sheet-form geometries, compared to floodplain and river terrace material, typified by well-rounded, better sorted, finer clasts. The stratigraphic interplay of the down-fan toe and the axial river floodplain or terrace is suggested as a key for improving our understanding of Quaternary fluvial evolutions triggered by tectonic or hydrological events (Mather et al. 2017).

The overbank, inactive portions of the fan surface undergo rapid dewatering and form a hardened surface with well-developed mud cracks that may be modified by aeolian deflation. A multitude of additional morphologies, which are not expressed by the resolution of topographic maps, can be found on a fan surface, among which are incised channels, telescopic lobs, stepped morphology, intersection points, gullies, bar-and-swale micromorphology, levees, fault scarps, boulder jams, wave-cut benches, and coastal bars.

15.1 Fan Terraces

A channel that sweeps laterally across the fan surface while incising forms *fan terraces*. The oldest fan surface is preserved as a terrace high up. Each successive lower terrace is younger. During a long term of entrenchment, short intervals of aggradation may cause *channel backfilling* by deposition of *insets*, i.e., younger deposits laid down in the entrenched channels. The inset fill may later become entrenched again and form an *inset terrace* (Fig. 15.1). A cycle of entrenchment, lateral migration, and re-entrenchment forms a *stepped fan morphology*. Channels that incise deep into the fan and continue into the bedrock form *strath terraces*, composed of the bedrock with an alluvial veneer (Srivastava and Misra 2008).

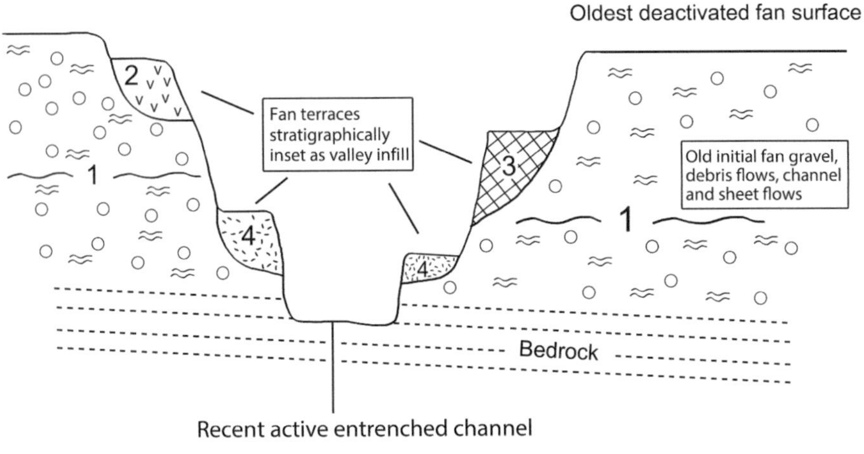

Fig. 15.1 Schematic profile across a channel displaying cut and fill

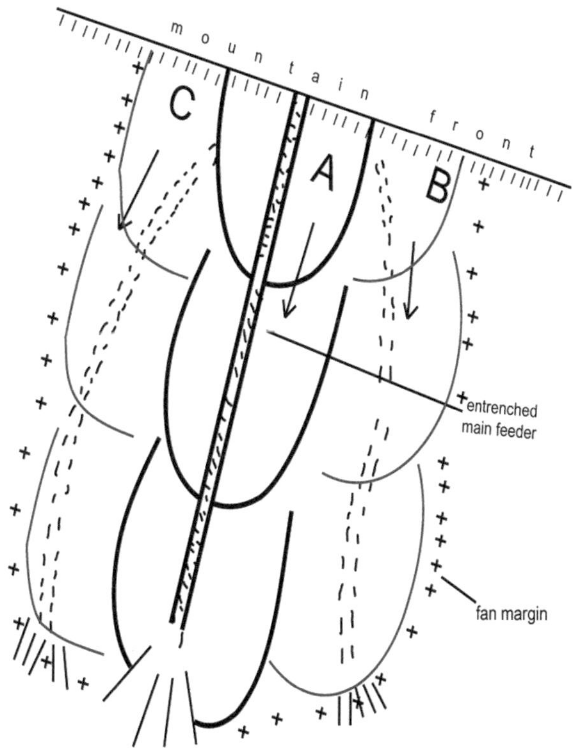

Fig. 15.2 Telescopic progradation and lateral migrations of lobes (**a**–**c**) which merge to build-up an alluvial fan (schematic). Lateral shifting among lobes will occur off and on in any stage

The longitudinal profiles of terraces *diverge* downslope in response to a falling base-level. Downslope *converging terraces* seem to indicate a gradual accommodation to a stable base-level. The fan terraces are not expected to persist upstream beyond the mountain front. When a sequence of lobes stretch down-fan, the fan surface exhibits a *telescopic pattern* (Fig. 15.2). Along the sequence, at the head of each lobe, the channel departs entrenched conditions at an intersection point. In the Dead Sea area, a telescopic sequence of 15 lobes which, following entrenchment, became terraces was formed during about 1000 years (Bowman 1978).

15.2 Surface Roughness

Roughness describes the elevation variability on the fan surface. *Grain roughness* is the smallest scale of roughness, caused by size differences between clasts (Amit and Gerson 1986; McFadden et al. 1989). Roughness on a larger scale describes the *microtopography* of the fan surface focused on the *bar-and-swale* relief measured by the height difference. Alluvial fans display an *internal roughness gradient*, i.e., the well-developed bar-and-swale morphology in proximal areas gradually dies away down-fan. The distal fan areas display lower degrees of roughness. In addition, we encounter roughness created by gullies and entrenched channels.

Surface Smoothing In addition to the *internal roughness gradient* fans are dominated by an overall, long-term *Surface Smoothing*. Long-term intense subaerial weathering, including salt splitting and granular disintegration, produces progressive in-situ clast breakdown evidenced by the decreasing size of the clasts and their increased angularity. Over time the clastic material, deposited mainly along bars by braiding channels, migrates from the bars into the swales. As fans age the bar-and-swale relief gradually decreases. Eventually, planation comes to an end by formation of a flat pavement surface with pebble coverage exceeding 90%.

The trend of decreasing roughness with increasing age suggests the roughness criteria as a measure of relative-age determination with surface smoothing occurring rapidly in the first thousand years. A bar-and-swale ghost morphology, in the form of pebble-concentrated patches indicating the trace of subdued depositional bars on a hyperarid (<50 mm) alluvial plain, abandoned at the latest 1.5–2.0 Ma, Pliocene to Early Pleistocene, has been reported by Matmon et al. (2009). Wells et al. (1987) suggested that from Late Pleistocene to Late Holocene, bar relief may decrease from decimeters to centimeters. The period from the abandonment of an active fan surface to its complete smoothing allows a very wide time range and is location-dependent. The smoothing trend can be reversed as a result of young channeling and headward tributary gully incision into old abandoned surfaces (Fig. 15.3) that *become progressively rougher with age*, i.e., roughness and age become positively correlated.

Degrees of roughness mirror the stages of the fan surface evolution, illustrated mainly after Frankel and Dolan (2007):

1. The most rugged bar-and-swale morphology occurs in active alluvial channels and on operating surfaces. Bars are built mostly of coarse cobbles and boulders, whereas swales consist of smaller pebbles and sand. No desert varnish, soil, or desert pavement are yet developed in this stage. Relief ranges from 0.5 to 1.5 m. The bar-and-swale morphology diminishes with decreasing clast size.
2. The next stage of surface evolution still displays a prominent bar-and-swale topography with a little varnish. The boulders and gravel disintegrate and slowly

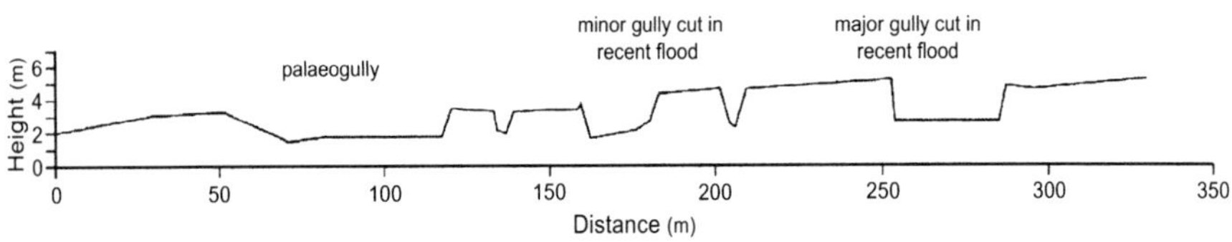

Fig. 15.3 Typical topographic profile across a fan showing gullies making a surface relief of 3 m. (Following McEwen et al. 2011)

decrease in size to be transported short distances from the bars into the adjacent swales by soil creep, slope wash, and rain splash.

3. As the fans mature, swales gradually fill in and become richer in fines. Remains of cobble bars still roughen the relief. The coarse gravels reside on the bars' surface longer than do finer sediments. An immature pavement starts to form, composed of finer-grained clasts with incipient soil formation, including Av and B horizons, with minor salt, carbonate, and clay-film accumulations. The relief becomes moderate, 0.25–0.75 m.
4. As fans age, the surface relief between bars and swales is subdued to 0.1–0.5 m, and the particle movement rate from bars to swales slows down. The bar-and-swale morphology gradually evolves toward a well-packed, heavily varnished pavement with a weakly developed C horizon.
5. The combined action of weathering and erosion leads to the final end product in the roughness evolution. The surface becomes a tightly packed, heavily varnished, mature and smooth desert pavement, with mature soil characterized by thick Av, Bt, and Bk horizons and with III–IV stages of carbonate development.
6. The trend of surface *entrenchment* takes over. The well-developed pavement becomes progressively dissected to form small patches divided by an entrenched channel network.

Establishing the Surface Roughness Roughness is shown in remote sensing through different brightness of the fan surface. The greater the roughness, the greater the scattering and reflectivity. Surface smoothness darkens the picture. The older the fan becomes, the darker the surface it shows. Characterization of alluvial surfaces includes visual interpretation of the surface brightness and hue on black and white and on color air photographs; spectral field measurements and airborne hyperspectral data analysis with accuracy assessment of the hyperspectral mapping (Crouvi et al. 2006). Mushkin and Gillespie (2005) showed that roughness can be estimated by measuring the ground-reflected solar radiance from the surface, dependent on the illumination and the angle view.

Application of airborne lidar-radar data to estimate the age of abandoned desert alluvial surfaces, based on the gradual smoothing process of the surface and on the capability of radar images to effectively quantify the roughness variability, has been demonstrated by Hetz et al. (2016). The method based on robust radar roughness data and on roughness-age correlations used surface ages based on previous luminescence and cosmogenic radionuclide datings. The average backscatter from a surface unit systematically decreases as a power-law function of time and can be effectively utilized as a proxy for surface age, i.e., inverted to radar-based age estimates ranging from Holocene to 1.6 Ma surfaces with uncertainty levels of ~35%. Later incision by rills on mature surfaces <100 ka in age was shown to *reverse* the age-progressive smoothing trajectory (Matmon et al. 2006; Guralnik et al. 2010). Another method for measuring surface roughness to distinguish different surface ages was by calculating the standard deviation of the surface slope in a 5 m^2 moving window, based on airborne laser swath mapping (Frankel and Dolan 2007).

15.3 Geomorphological Data

Characterization of the fan surface should include mapping the major geomorphic features while observing the postdepositional modifications. The mapping should include the accretionary and erosive geomorphic features, i.e., channels, levees, boulder fields, exposures, sieve deposits, lobes, debris plugs, terraces, intersection points, and islands. Cores should be collected from stratigraphic test holes and geophysical logs from wells.

Proximal-distal surface profiles provide the longitudinal slope gradients along the main active channel and along additional surfaces of the fan. Cross sections of channels and gullies and aerial photographs and satellite imagery bear witness of the form and the organization of the drainage system and of tectonic impacts, often clearly displayed by linear fault scarps. Mapping the contacts of coalescing fans can be effectively accomplished on enlarged aerial photographs used as base maps and later checked in the field. Landsat imagery at 15, 30, and 60 m resolution and panchromatic, multispectral, thermal, and DEM data are useful to analyze layers of slope, roughness, and vegetation type and density.

References

Amit R, Gerson R (1986) The evolution of Holocene Reg (gravelly) soils in deserts: an example from the Dead Sea region. Catena 13(1):59–79

Bowman D (1978) Determination of intersection points within a telescopic alluvial fan complex. Earth Surf Process 3:265–276

Crouvi O, Ben-Dor E, Beyth M, Avigad D, Amit R (2006) Quantitative mapping of arid alluvial fan surfaces using field spectrometer and hyperspectral remote sensing. Remote Sens Environ 104:1031

Frankel KL, Dolan JF (2007) Characterizing arid region alluvial fan surface roughness with airborne laser swath mapping digital topographic data. J Geophys Res 112:F02025. https://doi.org/10.1029/2006JF000644

Guralnik B, Matmon A, Avni Y, Fink D (2010) Be exposure ages of ancient desert pavements reveal quaternary evolution of the Dead Sea drainage basin and rift margin tilting. Earth Planet Sci Lett 290:132–141

Hetz G, Mushkin A, Blumberg DG, Baer G, Ginat H (2016) Estimating the age of desert alluvial surfaces with spaceborne radar data. Remote Sens Environ 184:288–301

Leeder MR, Mack GH (2001) Lateral erosion (toe cutting) of alluvial fans by axial rivers: implications for basin analysis and architecture. J Geol Soc Lond 158:885–893

Mather AE, Stokes M, Whitfield E (2017) River terraces and alluvial fans: the case for an integrated quaternary fluvial archive. Quat Sci Rev 166:74–90

Matmon A, Nichols K, Finkel R (2006) Isotopic insights into smoothening of abandoned fan surfaces, southern California. Quat Res 66:109–118

Matmon A, Simhai O, Amit R, Haviv I, Porat N, McDonald EV, Benedetti L, Finkel RC (2009) Desert pavement-coated surfaces in extreme deserts present the longest-lived landforms on Earth. Geol Soc Am Bull 121(5–6):688–697

McEwen LJ, Owen G, Matthews JA, Hiemstra JF (2011) Late Holocene development of a Norwegian alpine alluvial fan affected by proximal glacier variations, episodic distal undercutting, and colluvial activity. Geomorphology 127(3):198–215

Mcfadden LD, Ritter JB, Wells SG (1989) Use of multiparameter relative-age methods for age estimation and correlation of alluvial fan surfaces on a desert piedmont, eastern Mojave Desert, California. Quat Res 32:276–290

Mushkin A, Gillespie AR (2005) Estimating sub-pixel surface roughness using remotely sensed stereoscopic data. Remote Sens Environ 99:75–83

Srivastava P, Misra DK (2008) Morpho-sedimentary records of active tectonics at the Kameng river exit, NE Himalaya. Geomorphology 96:187–198

Wells SG, McFadden LD, Dohrenwend JC (1987) Influence of late quaternary climatic changes on geomorphic and pedogenic processes on a desert piedmont, eastern Mojave Desert, California. Quat Res 27(2):130–146

Dating of Alluvial Fans

16

Abstract

The age of a fan surface marks the end of the merging of lobes as part of the depositional fan buildup and the approximate time that elapsed since the start of trenching and abandonment of the fan surface. A wide scatter of ages over one fan results from the time needed to gradually tie lobes together into one alluvial fan unit. Most fan ages spread from younger than 10 K to Mid-Pleistocene. The erosional response of a source area to uplift and the subsequent progradation of gravel and deposition of a fan often lag geological time spans, even up to more than a million years, behind the uplift event.

Fan surfaces are dated by $_{230}Th/_{234}U$ of secondary $CaCO_3$ and by radiocarbon geochronology if datable organic material is available. Stabile surfaces are good candidates for the terrestrial cosmogenic nuclide method that measures the length of the time that a given geomorphic surface has been exposed to cosmic radiation since its deposition. In hyperarid regions where datable material, whether organic matter or pedogenic carbonate, is scarce, optically stimulated luminescence (OSL) dating is the efficient technique. Fan surfaces can as well be dated by using the rock varnish on the surface of clasts. Being time-transgressive, the fan surface provides useful features as pavement characteristics for estimating the relative fan age.

Keywords

Optically stimulated luminescence · Terrestrial cosmogenic nuclide method · Exposure period · Inheritance · Radiocarbon geochronology · Time indicator · Relative fan age · Equilibrium aggradation

Uplift and Progradation The erosional response of a source area to uplift will often lag behind the event, dependent on the lithology, on the structure of the bedrock, on the climate in the uplifted mountain block, and on the duration of the sediment

© Springer Nature B.V. 2019

D. Bowman, *Principles of Alluvial Fan Morphology*,
https://doi.org/10.1007/978-94-024-1558-2_16

transport through the fluvial system. Heller et al. (1988), Burbank et al. (1988), and Jones et al. (2004) analyzed the development of foreland basins, the progradation of the gravel, and the lag of the depositional phase. Carretier and Lucazeauw (2005) reported the lag of the maximum incision at the exit point of the mountains.

The time lag of arrival of a gravel front in the basin, expressed in the stratigraphy by a markedly thicker-bedded gravel unit, increases with distance. Such time span between the tectonic event and the arrival of the gravel front can reach geological scales on the order of more than a million years within as little as 20 km downstream from the thrust front. One needs to be aware of where stratigraphic data are collected with respect to the mountain front in order to interpret tectonic timing. Where the basins quickly became filled because of high sediment sourcing, as in the Himalayas, gravel can well serve as a syntectonic indicator of thrusting. Gravel may prograde after the start of thrusting a few tenths of km/my.

16.1 Absolute Dating

The age of the fan surface represents the approximate time elapsed since the last significant flooding and the start of trenching and abandonment. Dating desert pavements provides therefore the minimum age estimate (Gosse and Phillips 2001). Multiple ages from a single fan surface should yield a discrete and narrow age range which clusters to about 10–20% of the average and should be distinguishable without overlap by younger or older fan surfaces. A wide scatter of ages over one single fan surface results from the complex lateral channel migrations needed to tie patchy sediments as lobes, laterally together, into one alluvial fan, emphasizing that alluvial fans are in fact merged lobes, each with its period of activity, making the fan a *diachronous sedimentary body*. Samples across fans, which show quite different ages, reflect different depositional periods of lobes during the fan buildup and emphasize that no synchronicity should be expected within fans. Prior to sampling the conceptual depositional evolution of the fans should be established.

A review of dating methods and their limitations has been provided by Watchman and Twidale (2002). Fan surface dating can be applied by $_{230}Th/_{234}U$ on secondary $CaCO_3$ dense laminated coats and on secondary $CaCO_3$ layers in the soil matrix. Radiocarbon geochronology – ^{14}C – can often not be applied due to the paucity of datable organic material in arid alluvial fans. The oldest growth rings of trees on the fan surface provide an estimate of the minimum age of the fan, taking into account a time lag between the fan formation and shrub establishment. The date of organic matter *within* a fan unit does not provide the onset and the termination of a depositional event. The unit overlying dated material may lag behind if an erosional hiatus is shown.

The *terrestrial cosmogenic nuclide* – TCN – method measures the length of time that a given geomorphic surface has been exposed to cosmic radiation since its deposition (Cockburn and Summerfield 2004). Cosmogenic nuclides are rare radioactive isotopes created when high-energy cosmic rays interact with the nucleus of a solar system atom (Ivy-Ochs and Schaller 2010). Rocks initially contain no cosmogenic nuclides. Over time, cosmogenic nuclides accumulate in the rock, and their concentration is proportional to the time of surface exposure (Lal 1988). The penetration of cosmic rays decreases exponentially below the surface, and therefore the concentration of cosmogenic nuclides decreases with depth. Their lifetimes range from thousands to millions of years. The most useful for dating are the radionuclides ^{10}Be, ^{14}C, ^{26}AL, and ^{36}CL. Beryllium-10 is used most widely as it has the best-determined production rate and can be measured at low concentrations (Gosse et al. 2003).

When the fan surface is exhumed, the uppermost layers that contain the highest nuclide concentrations are lost, and new deeper surfaces are exposed. For dating purpose, surfaces should maintain a stable appearance that suffered the least erosion over the duration of the exposure to cosmic rays. Boulders located on flat portions of the fan surface, away from dissecting channels, especially when continuously exposed and undergone by minimal weathering, assure surface stability and are good candidates for fan surface dating. ^{10}Be concentrations in surficial boulders is directly related to the inactive "exposure period" postdating the abandonment of the fan surface following entrenchment. Lack of fan-head incision and continued fluvial activity prevent preservation of exposed conditions. Burial shows when nuclide production ceases (Ivy-Ochs et al. 2013).

Cosmogenic nuclides are particularly useful in regions that either lack datable organic matter or have ages beyond 45 kyr, the upper limit of radiocarbon dating, and beyond the several hundred-thousand-year range of the luminescence method. The cosmogenic nuclides may not have been produced exclusively on one respective fan surface but also during former sedimentary transport cycles of the clasts. The length of the exposure time can be grossly overestimated by the presence of such old clasts that carry evidence of *inheritance*. Boulders may thus yield ages that exceed the depositional age of the fan surface by tens thousands of years due to former transport histories and remobilization in the sourcing basin (Matmon et al. 2005; Schmidt et al. 2011).

In hyperarid regions where datable material, whether organic matter or pedogenic carbonate, is scarce, *optically stimulated luminescence* (OSL) dating is the recommended method for dating aggradational geomorphic surfaces (Porat et al. 2010). When minerals are buried and shielded from light, the impact of natural radioactivity builds up a *luminescence* signal, proportional to the time spent in darkness, which is used for dating. The signal is erased and the clock reset by exposure to daylight during erosion and transportation. Luminescence dating requires a proper erasure and resetting of the previously acquired luminescence clock to a very low level, either by exposure to sunlight or by a thermal event, otherwise the burial time will be added to previous burial times and the surface will appear older than it is in reality. Complete resetting of the OSL signal in the alluvial environment (i.e., bleaching) is rather an exception (Lang 2013) because exposure has often been insufficient to reduce the apparent dose to a negligible fraction of the final burial dose. Sediments that have undergone considerably longer and more numerous transport and deposition cycles, prior to final deposition, allow greater bleaching than modern sediments. Coarser grain sizes are better bleached probably because they are transported at a slower rate, allowing more time for sunlight exposure (Rittenour 2008). Following the zeroing event and upon the subsequent burial, the natural minerals begin luminescence acquisition afresh from the natural radioactive elements U238, Th232, and K40 present in the sediments. The time elapsed since deposition can be obtained by the ratio of the total measured luminescence acquired to the annual rate of acquisition (Aitken 1985, 1998).

The maximum age range of the optically stimulated luminescence dating technique is limited to about 300–350 ka (Murray and Olley 2002). The luminescence age of mineral grains underlying a fan surface represents their last exposure to sunlight before burial by the uppermost surface fan unit which is somewhat younger but close in time. The burial age of the uppermost beds is suggested as the proxy of the fan surface abandonment. Fan surfaces have been also dated by using the rock varnish on surface clasts (Dorn 1988).

Fan surfaces may persist over 10^5–10^6 years. Bull (1991, 1996) dated fan surfaces older than 1200 ka, that is, of Early-Middle Pleistocene age. The oldest alluvial fan with a well-developed gypsic horizon and advanced erosion, dated by Amit and Yaalon (1996), was 231+/− 22 ka in age. A bajada surface in the western central depression of the north Chile Coastal Cordillera appears by cosmogenic and luminescence dating to have been effectively fossilized at least since 230 ka (Dunai et al. 2005). Many of the fan chronologies were established on the 1000–100,000-year timescale. Many are younger than 10 k, that is, Holocene fans that postdated the final late glacial recession.

It is possible to make some estimates for the ages of fan surfaces, based on the pavement properties. Amit et al. (1993) and Amit and Gerson (1986) identified immature pavements on surfaces younger than ca. 14 ka and suggested approximate timescales of at least 100 ka for the development of a mature smooth pavement surface with fully shattered small clasts and with a thick pedogenic carbonate phase. On the other hand, active channel deposits are frequently several hundred to several thousand years old. In active channels, recently abandoned bar sediments carried ages of 500–5000 years (Porat et al. 2010).

16.2 Relative Dating

Being time-transgressive, i.e., changing with time, the fan surface provides useful characteristics for estimating the *relative surface age* (Mcfadden et al. 1989), provided they are similar in deposit particle size and in lithology. The following *time indicators* are recommended to estimate the relative surface age:

1. The elevation of the fan surface above the modern active stream channel (Bull 1991). The higher the terraces, the older and more mature the surface. Alluvial fan deposits that form isolated “hills” or “islands” high on a fan surface may designate residuals of older surfaces.

2. Rock varnish on clasts (Sowers 2000), ranging from none or incipient to well-developed. The changes in varnish cover are independent of deposit particle size. Reddening of clast and rock coating color can be determined by the Munsell color chart, and the differences in the surface brightness can be manifested in air photos.
3. The roughness of the fan surface, i.e., the degree by which the original bar and the interbar swale topography are well preserved or completely erased (Bull 1996).
4. The degree of chemical and mechanical weathering shown by the disintegration of surface clasts (Birkeland and Noller 2000). This ranges from unweathered and unfractured clasts to strongly fractured and pitted debris down to complete disintegration. The mean widths of boulder weathering fractures may strongly correlate with the alluvial fan surface age (D'arcy et al. 2015), suggesting that the time-integrated rate at which weathering fractures are widened in granitic surface boulders is 1.05 ± 0.03 mm ka^{-1} for the last 140 ka.
5. The abundance of the types and the magnitude of the diversity of vegetation which decreases at higher and older surfaces.
6. The degree of the soil maturity (Rockwell 2000) expressed by the soil profile development, including the thickness of the aeolian silt (Av), the development of the calcareous Bwk horizons, and the enrichment of gypsum.
7. The stage of desert pavement development within the array: without pavement, with immature pavement, moderately developed, mature, or well-developed pavement. Highly developed desert pavement exhibits high gravel coverage and well-sorted, small gravel size (Amit and Gerson 1986).
8. The degree of surface dissection, determined by visual estimation (Amit and Yaalon 1996).
9. The stage of fault scarp diffusion, often implied in paleoseismic studies (Hsu and Pelletier 2004). The method uses a diffusivity coefficient of the scarp surface k m^2/ky and can also be applied on risers of fan terraces and on scarps formed by gullies.
10. Field relation to well-dated features. Alluvial fans can be assigned approximate ages when covering, overlying, or interfingering with dated shoreline deposits, lake sediments, or paleosols. Shorelines may either cover older alluvial fans, or be covered by younger fans.

16.3 Rates of Deposition and Erosion

Swanson-Hysell and Barbeau (2007) reported Early to Late Oligocene sedimentation rates of ~32 cm/ka for alluvial fans, differentiated to ~16 or less cm/ka for the distal fan environment and 100 cm/ka in the proximal fan zone. These rates agree with recently reported vertical average aggradation rates of 0.01–0.46 mm/year. Higher extreme rates have also been reported. Sedimentation rates diminish as the fan grows. When the deposited sediment just balances the subsidence rate, fan elevation remains stable with no sediment left for progradation, and the fan enters a state of *equilibrium aggradation* (Parker et al. 1998).

Cosmogenic nuclide concentrations measured on abandoned fan surfaces along the Mojave section of the San Andreas Fault (Matmon et al. 2006) showed on bars of the fan surface average eroding rates of 27–29 mm kyr $^{-1}$. Similar erosion rates, of ~30 mm kyr^{-1}, were shown for Mid-Pleistocene terraces in humid France (Siame et al. 2004). These eroding rates of 0.02–0.03 mm/year are up to one scale order slower than the above mentioned sedimentation rates. Fan surface lowering rates decrease as fan ages increase. Slow erosional rates of 0.001 mm/year have been reported for glacial outwash surfaces in Patagonia (Hein et al. 2009) and for surfaces in the hyperarid Negev Desert (Matmon et al. 2009). In the Atacama Desert, the oldest alluvial fan surfaces typified by long surface exposure showed even more extremely low erosion rates of 0.0019–0.00003 mm/year derived from ^{10}Be exposure ages. These rates refer to relicts of Middle Miocene fossil alluvial remnants and probably possess the slowest erosion rates on earth (Nishiizumi et al. 2005). Long-term hyperaridity and geologic stability can preserve surface features for many millions of years.

References

Aitken MJ (1985) Thermoluminescence dating. Academic, New York, 359 pp

Aitken MJ (1998) An introduction to optical dating. University Press, Oxford, 267 pp

Amit R, Gerson R (1986) The evolution of Holocene Reg (gravelly) soils in deserts: an example from the Dead Sea region. Catena 13(1):59–79

Amit R, Yaalon DH (1996) The micromorphology of gypsum and halite in reg soils-the Negev Desert, Israel. Earth Surf Process Landf 21(12):1127–1143

Amit R, Gerson R, Yaalon DH (1993) Stages and rate of the gravel shattering process by salts in desert Reg soils. Geoderma 57(3):295–324

Birkeland PW, Noller JS (2000) Rock and mineral weathering. In: Quaternary geochronology: methods and applications. American Geophysical Union, Washington, DC, pp 293–312

Bull WB (1991) Geomorphic responses to climatic change, 326 p

Bull WB (1996) Global climate change and active tectonics: effective tools for teaching and research. Geomorphology 16(3):217–232

Burbank DW, Beck RA, Raynolds RGH, Hobbs R, Tahirkheli RAK (1988) Thrusting and gravel progradation in foreland basins: a test of post-thrusting gravel dispersal. Geology 16:1143–1146

Carretier S, Lucazeauw F (2005) How does alluvial sedimentation at range fronts modify the erosional dynamics of mountain catchments. Basin Res 17:361–381

Cockburn HAP, Summerfield MA (2004) Geomorphological applications of cosmogenic isotope analysis. Prog Phys Geogr 28(1):1–42

D'arcy M, Roda Boluda DC, Whittaker AC, Carpineti A (2015) Dating alluvial fan surfaces in Owens Valley, California, using weathering fractures in boulders. Earth Surf Process Landf 40(4):487–501

Dorn RI (1988) A rock-varnish interpretation of alluvial-fan development in Death Valley, California. Natl Geogr Res 4:56–73

Dunai TJ, Lopez GAG, Juez-Larre J (2005) Oligocene-miocene age of aridity in the Atacama Desert revealed by exposure dating of erosive sensitive landforms. Geology 33:321–324

Gosse JC, Phillips FM (2001) Terrestrial in situ cosmogenic nuclides; theory and application. Quat Sci Rev 29:1475–1560

Gosse J, McDonald E, Finkel R (2003) Cosmogenic nuclide dating of arid region alluvial fans. Geol Soc Am Abstr Programs 35(3):8

Hein AS, Hulton NRJ, Dunai TJ, Schnabel C, Kaplan MR, Naylor M, Xu S (2009) Middle Pleistocene glaciation in Patagonia dated by cosmogenic-nuclide measurements on outwash gravels. Earth Planet Sci Lett 286:184–197

Heller PL, Angevine CL, Winslow NS, Paola C (1988) Two-phase stratigraphic model of foreland basin development. Geology 16:501–550

Hsu L, Pelletier JD (2004) Correlation and dating of quaternary alluvial-fan surfaces using scarp diffusion. Geomorphology 60:319–335

Ivy-Ochs S, Schaller M (2010) Examining processes and rates of landscape change with cosmogenic radionuclides. In: Froehich K (ed) Environmental radionuclides – tracers and timers of terrestrial processes. Elsevier, Amsterdam, pp 232–294

Ivy-Ochs S, Dühnforth M, Densmore AL, Alfimov V (2013) Dating fan deposits with cosmogenic nuclides. In: Schneuwly-Bollschweiler M, Stoffel M, Rudolf-Miklau F (eds) Dating torrential processes on fans and cones, advances in global change research, vol 47. Springer, Dordrecht, pp 243–263

Jones MA, Heller PL, Roca E, Garces M, Cabrera L (2004) Time lag of syntectonic sedimentation across an alluvial basin: theory and example from the Ebro basin. Spain Basin Res 16:489–506

Lal D (1988) In-situ produced cosmogenic isotopes in terrestrial rocks. Ann Rev Earth Planet Science 16:355–388

Lang A (2013) Luminescence dating of alluvial fans and cones. In: Schneuwly-Bollschweiler M, Stoffel M, Rudolf-Miklau F (eds) Dating torrential processes on fans and cones. Advances in global change research, vol 47. Springer, Dordrecht, pp 283–295

Matmon A, Schwartz DP, Finkel R, Clemmens S, Hanks T (2005) Dating offset fans along the Mojave section of the San Andreas fault using cosmogenic ^{26}Al and ^{10}Be. GSA Bull 117:795–807

Matmon A, Nichols K, Finkel R (2006) Isotopic insights into smoothening of abandoned fan surfaces, Southern California. Quat Res 66:109–118

Matmon A, Simhai O, Amit R, Haviv I, Porat N, McDonald E, Benedetti L, Finkel R (2009) Desert pavement-coated surfaces in extreme deserts present the longest-lived landforms on Earth. Geol Soc Am Bull 121:688–697

Mcfadden LD, Ritter JB, Wells SG (1989) Use of multiparameter relative-age methods for age estimation and correlation of alluvial fan surfaces on a desert piedmont, eastern Mojave Desert, California. Quat Res 32:276–290

Murray AS, Olley JM (2002) Precision and accuracy in the optically stimulated luminescence dating of sedimentary quartz: a status review. Geochronometria 21:1–16

Nishiizumi K, Caffee MW, Finkel RC, Brimhall G, Mote T (2005) Remnants of a fossil alluvial fan landscape of Miocene age in the Atacama Desert of northern Chile using cosmogenic nuclide exposure age dating. Earth Planet Sci Lett 237:499–507

Parker G, Paola C, Whipple KX, Mohrig DC (1998) Alluvial fans formed by channelised fluvial and sheet flow, I. Theory. J Hydraul Eng 124:985–995

Porat N, Amit R, Enzel Y, Zilberman E, Avni Y, Ginat H, Gluck D (2010) Abandonment ages of alluvial landforms in the hyperarid Negev determined by luminescence dating. J Arid Environ 74:861–869

Rittenour TM (2008) Luminescence dating of fluvial deposits: applications to geomorphic, palaeoseismic and archaeological research. Boreas 37:613–635

Rockwell TK (2000) Use of soil geomorphology in fault studies. In: Quaternary geochronology: methods and applications. American Geophysical Union, Washington, DC, pp 273–292

Schmidt S, Hetzel R, Kuhlmann J, Mingorance F, Ramos VA (2011) A note of caution on the use of boulders for exposure dating of depositional surfaces. Earth Planet Sci Lett 302:60–70

Siame L, Bellier O, Braucher R, Sébrier M, Cushing M, Bourlès D, Hamelin B, Baroux E, de Voogd B, Raisbeck G, Yiou F (2004) Local erosion rates versus active tectonics: cosmic ray exposure modelling in provence (south-east France). Earth Planet Sci Lett 220:345–364

Sowers JM (2000) Rock varnish chronometry. In: Quaternary geochronology: methods and applications. American Geophysical Union, Washington, DC, pp 241–260
Swanson-Hysell N, Barbeau DL (2007) The diachroneity of alluvial-fan lithostratigraphy? A test case from southeastern Ebro basin magnetostratigraphy. Earth Planet Sci Lett 262(3):343–362
Watchman AL, Twidale CR (2002) Relative and "absolute" dating of land surfaces. Earth Sci Rev 58:1–49

The System Approach

17

Abstract
Alluvial fans do not create steady and permanent landforms but are a transient feature with alternating phases of aggradation and entrenchment. The fans may reach a dynamic equilibrium when the supply of material from the source catchment is balanced by erosion, and when the subsidence rate, the fan elevation, and the shape remain constant. This is, however, a rare situation. Alluvial fans compose, more commonly, a non-equilibrium system and are hardly expected to reach stability. The evolution of alluvial fans involves both negative and positive self-enhancing feedbacks. Aggradation is a self-arresting feedback mechanism. Oversteepening following aggradation induces trenching and gullying as a negative feedback which leads to a reduction of the channel slope and to a more equilibrated grade, moving the stream power/resisting power ratio closer to the value of 1.0 and to improved coupling. Through-fan-trenching marks the ultimate stage of coupling, allowing continuity of the sediment movement through the fan into the axial basin. The alluvial fans are controlled by both autogenic and allogenic forcing, the latter not always necessary to explain significant changes.

Keywords
Transitional phase · Steady state · Dynamic equilibrium · Positive self-enhancing feedback · Negative self-arresting feedback · Intrinsic control · External forcing · Coupling · Aging

17.1 Transitional Unstable Conditions

The evolution of fans includes the building of a depositional relief and its subsequent destruction as part of a morphological cycle, shaped by both deposition and erosion. Fans show a relatively short-lived, active morphology, which becomes gradually consumed with time. As alluvial fans typically increase in volume and

© Springer Nature B.V. 2019
D. Bowman, *Principles of Alluvial Fan Morphology*,
https://doi.org/10.1007/978-94-024-1558-2_17

later degrade and loose sediments, they constitute a *transitional* sedimentary body. Most fans are prone to frequent climatic and tectonic perturbations by incessant alternations of aggradation and entrenchment which counteract steady state and do not allow the time needed to reach equilibrium. In alluvial fans lobes keep prograding down-fan, intersection points shift and avulsion enhances lateral trends. Hence, fans are not sites that promote equilibrium nor stability. When the volume of sediments accumulating on a fan equals the volume removed from it, the fan system approaches a certain steady state. A fan in a graded state should also preserve its dimension and geometry. However, such dynamic equilibrium is not common.

17.2 Feedbacks

Evolution of alluvial fans involves both negative and positive feedbacks. Bull (1979) used the term *self-enhancing processes* to define a regime that drives processes toward a peak. An example of such a process is the deposition along the down-fan profile which is enhanced by the decreasing discharge owing to infiltration, by the diminishing slope grade, by the lateral spread of the braided channels, and by the sheet flows that decrease the flow depth (Bull 1997), reduce the carrying capacity, and boost aggradation. Another example is the removal of soil from slopes in sourcing basins which increases outcrop areas that further intensify runoff and entrenchment as a self-enhancing process.

Logjams that inhibit transport are an example of a negative, self-arresting feedback mechanism by promoting counterbalancing trends: logs or other floating coarse vegetation debris decrease the channel gradient and slow the channel flow along short channel segments, counteracting incision and even changing the trend into deposition. An additional example are fans with higher rates of deposition and thickening that encroach upon their neighbors and as a result will decrease their own rate of increase in thickness, thus demonstrating a negative feedback (Fig. 17.1). The encroached fans will gradually thicken, and the negative feedback may finalize in a quasi-equilibrium.

When alluvial fans grow and gradually create an oversteepened fan head which exceeds the *stability threshold* (Bull 1997), a negative feedback in the form of

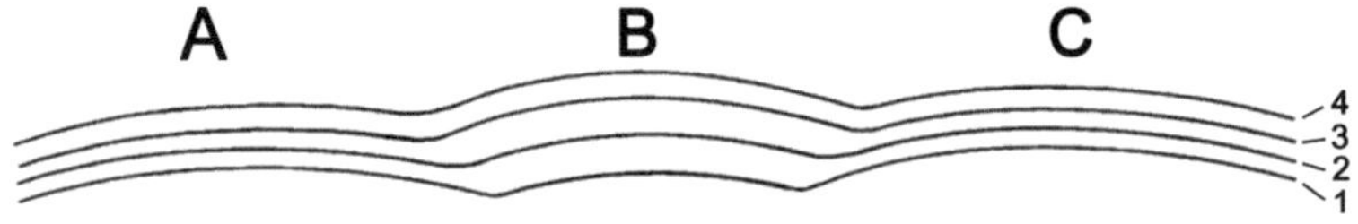

Fig. 17.1 Hypothetical transverse sections across adjacent coalescing fans. The higher rate of deposition on fan B triggers lateral encroachment on fans A and C and gradually decreases its own rate of thickening. Being encroached, fans A and C become narrower and increase their rate of thickening. The different trends finally reach a steady state in form of a similar rate of deposition (lines 3–4) and stabilized boundaries. (After Hooke 1968)

downcutting and entrenchment will be initiated (Bull 1979) and reduce the channel slope.

17.3 Controls

A distinction should be made between *intrinsic/autogenic dynamics* and *allogenic/external forcing*. Autogenic refers to a morphologic change that arises from the alluvial fan's internal dynamics like changes in the hydraulic geometry or changes in flow direction, as opposed to allogenic changes that result from external forcing, such as tectonic uplift, climate change, anthropogenic activity, and, in littoral environments, a falling or rising sea/lake level.

Intrinsic/Autogenic Forcing Allogenic forcing is not always the reason for significant landscape changes. Much of the variability is autogenic (Paola et al. 2009). Sedimentary cycles of accumulation/degradation may be entirely due to intrinsic processes which do not require allogenic controls and can result, for example, of oversteepening and exceedance of an intrinsic threshold slope. *Geomorphic intrinsic thresholds*, when exceeded, start a significant landform change inherent in the normal evolution without extrinsic changes. Crossing thresholds play a prominent role as intrinsic forcing (Schumm 1973, 1979). Fans should not be approached as a homogeneous sedimentary body, and threshold effects should not be expected to be uniformly distributed across it.

Reduced sediment availability with time, i.e., aging (Eckis 1928), is an intrinsic development leading to growing fan dissection. Migration of alluvial channels by avulsion is also a typical feature of autogenic behavior. An avulsion event is estimated to occur on an order of 10^2–10^4 years timescale (Heward 1978). Localized fan degradation, with no wide lateral spread nor synchronicity with adjacent fans, suggests an autogenic control. Differing fan responses in similar regional conditions of climate may too be caused by intrinsic processes (Harvey 1997). A regional similarity in the behavior of fans may suggest that the controlling factors are not intrinsic.

Allogenic/External Forcing A regional similarity in the behavior of fans points to allogenic/external changes that are driven by factors such as human land use, changes in water and sediment supply rates caused by climate, base-level change, and tectonic activity. External changes have greater spatial and temporal effects and include long-term cycles that last 10^5–10^6 years (Heward 1978). The introduction of external controlling factors displays the alluvial fans as open systems. External environmental forcing is, however, not always necessary even for explaining dramatic and persistent landscape changes (Nicholas and Quine 2007), which may result entirely from internal dynamics.

17.4 Coupling

Coupling relates to the connectivity within the fluvial system (Harvey, 1987, 1997, 2002, 2012). Alluvial fans act as buffers, i.e., trapping sediments derived from a mountain source, and thus become a storage which causes *disconnectivity* between the source area and the axial terminal sink (Fig. 1.1). Aggrading alluvial fans isolate lowlands from the source of the sediment supply. Where the system is buffered by alluvial fans, only fine materials may be available to the axial river.

Aggradation and buffering continue during the partial dissection phases, when erosion is restricted to the fan head or mid-fan and the zone of sedimentation is on the fan without leakage to the main downstream axial basin or to a playa. Under such conditions the drainage basin with the alluvial fan makes up a *closed system*. Coupling improves as the alluvial fan becomes more and more entrenched. A perfect coupling means that there is transmission of sediment through the entire system, from the headwater catchment area through the alluvial fan down to the axial basin. This is the stage of *through-fan trenching*, i.e., a complete crossing and bypassing of the fan, allowing continuity of sediment movement down into the basin. Major flow events and humid climates tend to intensify entrenchment and improve coupling.

References

Bull WB (1979) Threshold of critical power in streams. Geol Soc Am Bull 90:453–464

Bull WB (1997) Discontinuous ephemeral streams. Geomorphology 19:227–276

Eckis R (1928) Alluvial fans in Cucamonga district, southern California. J Geol 36:111–141

Harvey AM (1987) Alluvial fan dissection: relationships between morphology and sedimentation. In: Frostick LE, Reid I (eds) Desert sediments: ancient and modern, Geological Society Special Publication, vol 35. Blackwell, Oxford, pp 87–103

Harvey AM (1997) The role of alluvial fans in arid zone fluvial systems. In: Thomas DSG (ed) Arid zone geomorphology: process, form and change in drylands, 2nd edn. Wiley, Chichester, pp 231–259

Harvey AM (2002) Effective timescales of coupling within fluvial systems. Geomorphology 44:175–201

Harvey AM (2012) The coupling status of alluvial fans and debris cones: a review and synthesis. Earth Surf Process Landf 37:64–76

Heward AP (1978) Alluvial fan and lacustrine sediments from the Stephanian A and B (La Magdalena, Cinera-Matallana and Sabero) coalfieds, northern Spain. Sedimentology 25:451–488

Hooke RL (1968) Steady-state relationships on arid-region alluvial fans in closed basins. Am J Sci 266:609–629

Nicholas AP, Quine TA (2007) Modeling alluvial landform change in the absence of external environmental forcing. Geology 35:527–530

Paola C, Straub K, Mohrig D, Reinhardt L (2009) The "unreasonable effectiveness" of stratigraphic and geomorphic experiments. Earth Sci Rev 97:1–43

Schumm SA (1973) Geomorphic thresholds and the complex response of drainage systems. In: Morisawa M (ed) Fluvial geomorphology, Publication in Geomorphology. State University of New York, Binghampton, pp 299–310

Schumm SA (1979) Geomorphic thresholds: the concept and its applications. Trans Inst Br Geogr 4:485–515

18 Natural Hazards

Abstract

Alluvial fans are sites associated with hazards and risks and demand hazard and risk assessment as a basic stage in planning and setting insurance rates. Typical hazards on alluvial fans include earthquakes, faulting, burial under mud and boulders, scour and undermining, subsidence, and swinging of flow paths. In controlling hazards, the watershed is as important as the fan. Hazard mapping should include, among others, potential avulsion routes, density of the vegetation, land use, deforestation, overgrazing, bedrock, landslide-prone terrains, erodible-stored sediments, rainfall-correlated runoff hydrographs, sediment discharge, history of forest fires, and long-term record of flood inundations. A probabilistic flood hazard map should predict areas of flooding for the 10-year and 100-year maximum floods.

Keywords

100-year flood · Hazard mapping · Hazard classes · Control dams · Managed local base-level · Forest fire

18.1 The Hazards

Alluvial fans are appealing sites for settlement along mountain fronts, offering scenic cooler outlays and connectivity with main transportation routes along axial valleys. Alluvial fans are thus marked by considerable urbanization. However, alluvial fans along mountain fronts are also sites of specific hazards and potential risks. A *hazard* is defined as a potentially damaging harmful event. The *risk* relates the hazard to its adverse consequences – damage or loss – through the exposure, the vulnerability, and the worth of the affected areas while regarding the probability of the occurrence. Typical hazards on alluvial fans include *earthquakes*, *faulting*, *burial under mud and boulders*, *scour and undermining*, *subsidence*, *flooding*, *and flow*

© Springer Nature B.V. 2019
D. Bowman, *Principles of Alluvial Fan Morphology*,
https://doi.org/10.1007/978-94-024-1558-2_18

path uncertainty (FEMA 1989). Damaging events may last minutes to hours, with decades to centuries of long calm periods in-between.

The following hazards along mountain fronts are the most frequent and/or the most destructive:

I. *Debris flows* possess the greatest potential for destruction of property and for loss of lives because of their high viscosity. Debris flows are generated in small, several km^2 watersheds, 15–25° steep, draining into short, relatively steep fans with a response time shorter than 1 h (Santangelo et al. 2012). Clear-cut logging and road building in steep areas trigger landslides which source debris flows. The movement of the front of a debris flow which is a few meters high and includes boulders a few meters in diameter carries a violent impact proportional to their unit weight and to the mean velocity squared. Debris flows also magnify the danger of avulsion: arrested debris flows block channels and divert the streamflow (Kellerhals and Church 1990). Fans of more than one debris flow per year are *high-frequency debris-flow fans.*
II. Channel avulsion and lateral diversion of floods are a traditional hazard. When unable to pass the discharge within the banks and overflowing, floodwater moves as an unconfined sheet flow which may incise new channels and change the flow path pattern. In August 2008 an avulsion relocated the Kosi River in India by about 60 km to the east in the central part of the megafan, and a major floodwater channel started flowing radially, endangering hundreds of villages, destroying croplands and bringing disaster to thousands of people.
III. Ruptures by earthquakes and faulting activity are frequent along mountain fronts.
IV. Water withdrawal from fan reservoirs causes emptying part of the former water pores, increasing the squeezing of the grains and leading up to a few meters of land subsidence. Enhancing recharge of permeable alluvial fans is thus of great interest.

18.2 Hazard Mapping

Alluvial fans as a land of developing towns and villages demand hazard and risk assessment as basic requirements in planning and for setting flood insurance rates (FEMA 1989, 1995; Schumm et al. 1996; Wilford et al. 2009). Two of the main questions relevant to flood insurance programs are: which parts of an alluvial fan are subject to flooding and, in order to prevent building on possible future drainage paths, what are the potential avulsion routes. It is recommended as well to classify the fan type whether depositional or entrenched and predict zones of future scour and of high rates of sediment deposition (Fuller 2013). The main damage on depositional fans will be concentrated in the apical to the mid-fan zone, whereas on entrenched fans flood-prone areas are located along shallow entrenched channels down the mid-fan and along the distal zone. The ability to predict and estimate the uncertainty of the flow path and the lateral spread of future flows is essential in selecting and implementing mitigation.

Mapping Details Hazard assessment requires mapping (FEMA 2000, 2004), starting in the watershed which is as important as the fan. Mapping should include the *density of the vegetation* and the *land use* that are most relevant to the surface flow volume. *Deforestation* and *overgrazing* by introducing large herds of livestock will enhance the sediment output. The recent conditions of sediment supply should be recognized in the sourcing catchment by *bedrock mapping* as well as by mapping the total *erodible sediment stored* and available in the drainage basin (Schneuwly-Bollschweiler et al. 2013). *Landslide-prone terrains* should be mapped as well, including *clear-cut logging areas*, *steep slope gradients*, *vegetation-free areas*, and *gullied slopes*. In each catchment a *debris flow assessment* should be carried out based on the physical basin characteristics and the available material in the foreseeable future. Probability should also be estimated based on the recent debris flow history. Knowledge on frequency and magnitude of events should be established on analysis of reports, testimonies, and chronicles of past catastrophes. Debris flow events can be dated by tree injuries. Growth disturbances in tree-ring series can be used to date debris flow events. Scars eventually heal by peripheral or callous growth around the scar. The year of the injury can be closely approximated by dating the first annual growth ring on the outside of the scar (Sigafoos 1964). *Rainfall* as a major triggering factor can be shown by the maximum annual cumulative amount. In addition, data is needed to evaluate the *runoff production rate* based on rainfall-correlated runoff hydrographs in former events. Mapping should also include warning signs of instability as *soil slips* on steep terrain and minor features such as *cracks and small scarps*.

Surficial morphological mapping includes features like levees, channel depths, lobes, and aggrading channels where banks are low or nonexistent, especially on the outside bends that are prone to avulsion and diffusive sheet-flow zones. Active and inactive surfaces should be mapped in detail and contain their altimetric separation. Confining the specific passes through which debris surges and floods may travel due to flow blockage is important as well as mapping structures that will withstand erosion and their effect on the flow path. Hazard assessment needs also historic long-term records of flood inundations with dates and magnitudes and damage and areal extension including historical aerial photographs.

Floods Maximum probable peak flows with water and sediment discharge, and estimates of flood depths and velocities is fundamental data needed for risk evaluation (Schick et al. 1999). The existing channels on fans cannot be relied on to convey the 100-year peak flow. The National Flood Insurance Program (NFIP) in the USA requires that areas that are subject to flooding by the 100-year flood, i.e., a hazard of a 100-year recurrence interval and a 1% probability in any year, as well as by the 10-year maximum flood (Pelletier et al. 2005), should be mapped and recommendations for buildable and non-buildable zones should be provided, an aim which seems difficult to attain (Cazanacli et al. 2002).

The following *main flood hazard classes* are distinguished as follows (House et al. 2010): *Very high flood hazard areas* are those with the well-defined active channels. *Moderate flood hazard areas* are characterized by infrequent activity during floods. *Low flood hazard areas* are the stable alluvial surfaces excluded from activity due to long-term deep fan trenching. When regarding the *magnitude-frequency* dilemma (Wolman and Miller 1960), there is no data available to suggest that either moderate frequent events or the high-magnitude low-frequency extreme floods are the dominant processes controlling alluvial fans. Research provides examples of both (Beaty 1974; Wells and Harvey 1987).

Forest Fires may become a major factor that contributes to hazards. Fires destroy vegetation, decay the roots anchoring the colluvium, and accelerate sediment supply from drainage basins to alluvial fans. Consumption of the tree canopy, the shrubs, and the soil-mantling litter reduces the rainfall interception and the storage rates. For several years following a fire, runoff rates often more than double due to removal of the vegetation and the litter cover. Simultaneously, the burned surfaces provide abundant loose, friable soil, and fine wood ash to the surface runoff (Meyer and Wells 1997; Oakley et al. 2017). When *postfire rain events* of sufficient intensity and duration strike a burned area, large pulses of sediment removal are produced, and aggradation rates are greatest (Benda 1990; Cannon et al. 1998; Frechette and Meyer 2009). Erosion yields or rates, in terms of mass per unit area per unit time, have been documented to increase dramatically, tens to hundred times, following wildfires. This contribution of fine sediments removed from burned slopes is important in generating debris flows (Harvey 1984; Oakley et al. 2017; Wells and Harvey 1987). Later, as the loose upper soil has been stripped away, the surface becomes subject to compaction by raindrop impact and sealed (Romkens et al. 1990). The rapid change toward a more compact, erosionally resistant substrate greatly reduces the fine sediment input to runoff, even before slopes become revegetated.

18.3 Protective Devices

Following hazard zoning, control countermeasures are invited with respect to past debris flow and flood-computed recurrence frequencies (Jackson 1987). The following protective devices may be variously applied:

(A) The basic measure to deal with flood hazards is to dam the channel further upstream (French 1987). Such *control dams* create reservoirs which reduce flow and intercept sediment transport, serving as temporary storage basins also for debris flows. The required capacity of such protective reservoirs is of high importance. Fine sediment accumulation seals the reservoir bottom and decreases their capacity with time. Loss of storage increases the probability of a subsequent breach by overflow. Breaching of a dam may cause catastrophic floods, bigger and more hazardous than the natural flood event. The reservoirs, including saddle topography fields, can serve as retention ponds for redirecting the water for irrigation.

(B) Deepening and straightening the channels and enlarging their cross section will increase their transport capacity (Kellerhals and Church 1990).

(C) Roads and culverts should be raised on the fan surface without excavation in order to prevent the usual intervention in form of a *managed local base-level*, i.e., an excavated channel which concentrates flow from the upfan toward a culvert that conveys the water under the road (Florsheim et al. 2001) and initiates headcutting upfan. Road construction should minimize flow interception and interference with the existing drainage routes. When planning roads on fans, the different fan morphologies, i.e., fossil surfaces, areas of active deposition and of entrenchment should be considered, and peak flows should be estimated (French 1992).

(D) For urban protection, deflecting dykes can totally prevent the entry of floodwater and divert it laterally around the town. The encircling floodways require an estimate of the maximum probable flood, both in terms of peak discharge and total sediment yield. The sediment quantities during floods are almost always beyond the capacity that the circumferential floodways are designed to handle. Storm flow conveyors should therefore be wider and deeper than the original natural alluvial fan distributary system. In hyperarid environments, a sediment concentration of 5–10% by weight, most of it bed material, can serve as a rule of thumb for the design of diversion ditches (Schick et al. 1997). The streets in fan cities and drainage ditches can be planned as water conveyors, i.e., making out of the roads an efficient urban drainage system including the underground sewer system (Schick et al. 1999). Reinforcement of foundations and upfan walls of buildings for protection against the moving water is always recommended.

Lateral mobility of channels on fans, which does not count for incised stable channels, is as important as the flood extent in determining what areas are at risk. There is no assumption of complete uncertainty of the behavior of channels on fans. Channels do not move across the fan surface at random, and in the long run each point on the fan has no equal probability of being flooded. The wider the fan's opening angle, the smaller the flood probability of each location on the fan, and as the length of the down-fan distance increases, so does the flood probability (Dawdy 1979). In addition, Bull (1964) showed a predominance of channels within 30° of the fan's medial axis, which is the preferred path of flow and thus the more dangerous area (French et al. 2001). Hazard is expected to decrease as the point of interest moves away from the medial radial zone. Flows are often also arrested by man-made structures such as roads that become responsible for a specific lateral flow spread (Rickenmann and Schiedl 2013). The spatial morphological activity on fans confirms that the proximal zone is the most frequently activated, even by moderate and smaller events. Risks thus decrease from the fan's proximal zone downward. Distal locations are affected mainly by big and extreme events and are spared the activity that affects only the higher fan zone.

References

Beaty CB (1974) Debris flow, alluvial fans and a revitalized catastrophism. Z Geomorphol Suppl 21:38–51

Benda L (1990) The influence of debris flows on channels and valley floors in the Oregon coast range, USA. Earth Surf Process Landf 15:457–466

Bull WB (1964) Geomorphology of segmented alluvial fans in western Fresno County, California. Prof Pap USGS 352-E:89–129

Cannon SH, Powers PS, Savage WZ (1998) Fire-related hyperconcentrated and debris flows on Storm King Mountain, Glenwood Springs, Colorado, USA. Environ Geol 35(2–3):210–218

Cazanacli D, Paola C, Parker G (2002) Experimental steep, braided flow: application to flooding risk on fans. J Hydraul Eng 128(3):322–330

Dawdy DR (1979) Flood frequency estimates on alluvial fans. J Hydraul Div ASCE 105(HY11):1407–1413

Federal Emergency Management Agency FEMA (1989) Alluvial fans: hazards and management. Doc. no. 165. Washington, DC

Federal Emergency Management Agency FEMA (1995) Appendix 5: studies of alluvial fan flooding. Guidelines and specifications. Doc. no. 37. Washington, DC

Federal Emergency Management Agency FEMA (2000) Guidelines for determining flood hazards on alluvial fans

Federal Emergency Management Agency FEMA (2004). Flood hazard mapping – flood hazard zone designations

Florsheim JL, Mount JF, Rutten LT (2001) Effect of base level change on floodplain and fan sediment storage and ephemeral tributary channel morphology, Navarro River, California. Earth Surf Process Landf 26:219–232

Frechette JD, Meyer GA (2009) Holocene fire-related alluvial-fan deposition and climate in ponderosa pine and mixed-conifer forests, Sacramento Mountains, New Mexico, USA. The Holocene 19(4):639–651

French RH (1987) Flood hazard assessment on alluvial fans: an examination of the methodology. In: Application of frequency and risk in water resources. Springer, Netherlands, pp 361–375

French RH (1992) Design of flood protection for transportation alignments on alluvial fans. J Irrig Drain Eng 118(2):320–330

French RH, Miller JJ, Curtis S (2001) Estimating the depth of deposition (erosion) at slope transitions on alluvial fans. J Hydraul Eng 127(9):780–782

Fuller JE (2013) Gaps in FEMA guidance for delineating flood hazards on active alluvial fans. J flood Eng 4(1):29–38

Harvey AM (1984) Debris flows and fluvial deposits in Spanish quaternary alluvial fans: implications for fan morphology. In: Koster EH, Steel RJ (eds) Sedimentology of gravels and conglomerates. Canadian Society of Petroleum Geologists, Memoir 10:123–132

House PK, Buck BJ, Ramelli AR (2010) Geologic assessment of piedmont and playa flood hazards in the Ivanpah Valley Area. Clark County, Nevada, Report 53. Nevada Bureau of Mines and Geology, University of Nevada, Reno

Jackson LE (1987) Debris flow hazard in the Canadian Rocky Mountains. Geological Survey of Canada, Ottawa, pp 86–11

Kellerhals R, Church M (1990) Hazard management on fans, with examples from British Columbia. In: Rachocki AH, Church M (eds) Alluvial fans a field approach. Willey, Chichester, pp 335–354

Meyer GA, Wells SG (1997) Fire-related sedimentation events on alluvial fans, Yellowstone National Park USA. J Sediment Res 67:776–791

Oakley NS, Lancaster JT, Kaplan ML, Ralph FM (2017) Synoptic conditions associated with cool season post-fire debris flows in the Transverse Ranges of southern California. Nat Hazards 88(1):327–354

Pelletier JD, Mayer L, Pearthree PA, House PK, Demsey KA, Klawon JE, Vincent KR (2005) An integrated approach to flood hazard assessment on alluvial fans using numerical modeling, field mapping and remote sensing. Geol Soc Am Bull 117(9–10):1167–1180

Rickenmann D, Scheidl C (2013) Debris-flow runout and deposition on the fan. In: Schneuwly-Bollschweiler M, Stoffel M, Rudolf-Miklau F (eds) Dating torrential processes on fans and cones. Springer, Netherlands, pp 75–93

Romkens MJM, Prasard SN, Whisler FD (1990) Surface sealing and infiltration. In: Anderson MG, Burt TP (eds) Process studies in hillslope hydrology, chapter 5. Wiley, New York

Santangelo N, Daunis-i-Estadella J, Di Crescenzo G, Di Donato V, Faillace P, Martin-Fernandez JA, Romano P, Santo A, Scorpio V (2012) Topographic predictors of susceptibility to alluvial fan flooding, Southern Apennines. Earth Surf Process Landf 37:803–817

Schick AP, Grodek T, Lekach J (1997) Sediment management and flood protection of desert towns: Effects of small catchments. In: Walling DE, Probst JL (eds) Human impact on erosion and sedimentation. Int Assoc Hydr Sci Publ 245: p 183–189

Schick AP, Grodek T, Wolman MG (1999) Hydrologic processes and geomorphic constraints on urbanization of alluvial fan slopes. Geomorphology 31:325–335

Schneuwly-Bollschweiler M, Stoffel M, Rudolf-Miklau F (eds) (2013) Dating torrential processes on fans and cones, methods and their application for hazard and risk assessment, Advances in Global Change Research, vol 47. Springer, Dordrecht

Schumm SA, Baker VR, Bowker MF, Dixon JR, Dunne T, Hamilton D, Merritts D (1996) Alluvial fan flooding. National Research Council (NRC), Commission on Geosciences, Environment and Resources, National Academy Press, Washington, DC 182 p

Sigafoos RS (1964) Botanical evidence of floods and floodplain deposition. U.S. Geological Survey, Professional Paper, p 485-A

Wells SG, Harvey AM (1987) Sedimentologic and geomorphic variations in storm generated alluvial fans, Howgill Fells, northwest England. Geol Soc Am Bull 98:182–190

Wilford DJ, Sakals ME, Grainger WW, Millard TH, Giles TR (2009) Managing forested watersheds for hydrogeomorphic risks on fans. B.C. Min. For. Range, For Sci Prog, Victoria, B.C. Land Manag. Handb. 61 p

Wolman MG, Miller JP (1960) Magnitude and frequency of forces in geomorphic processes. J Geol 68:54–74

The Regional Approach: Alluvial Fans along the Dead Sea-Arava Rift Valley

19

Abstract

This review focuses on the rift valley between the Dead Sea and the Gulf of Elat, Israel, along part of the *Dead Sea Transform* that stretches north to south from Syria through the Red Sea to Africa. The transform has been intermittently active since at least the Miocene and remained the center of seismic activity, experiencing destructive earthquakes to the present day. The main issues of fan research at the Dead Sea area are fan morphology and the effect of a lowering base-level. Southward, along the Arava Rift Valley, alluvial fans have been studied mainly as surfaces that preserve tectonic fingerprints and allow dating normal and sinistral tectonic slips.

Various kinds of fans are exposed in the Dead Sea area: sublacustrine fans, Gilbert-type fans, fan deltas, and alluvial fans, often of the telescopic type. The drop of Lake Lisan and the Dead Sea provided the altitudinal range needed for intensive entrenchment and formed alluvial fans that are unique by their deep incision, high number of fan terraces, and raised shorelines. There are, however, major reasons to deny an active lowering-base-level control in the fan environment along the Dead Sea. The rapid lake level drop only allowed entrenchment but did not cause it. Fan entrenchment and progradation have been activated from upstream, i.e., a "top-down" driven entrenchment while lagging behind the rapid lake level drop.

Southward, along the Arava Rift margins, fans are transected and displaced by normal and strike-slip faults of the Dead Sea Transform system. Alluvial fans show sharp linear boundaries, and some are offset horizontally relative to their feeding catchment. The calculated horizontal slip rates are consistent with the present-day GPS slip.

Keywords

Dead Sea Transform · Lake Lisan · Lisan Formation · Sublacustrine fan · Fan terrace · Beach terrace · Gilbert-type fan · Fan delta · Telescopic fan · Base-level effect

© Springer Nature B.V. 2019
D. Bowman, *Principles of Alluvial Fan Morphology*,
https://doi.org/10.1007/978-94-024-1558-2_19

Being sensitive recorders, alluvial fans provide information about events in their sourcing basins as well as in the accommodation area. As the response to climate or to tectonic activity is likely to be regionally wide and synchronous, the alluvial fans constrain the age of regional depositional or incisional events (Suresh et al. 2002) and preserve evidence related to the regional faulting history including neotectonic activity.

Regional-focused studies related to alluvial fans have been performed, for example, in the southeast USA (Harvey et al. 1999; Sohn et al. 2007), in southeast Spain (Harvey 1990; Silva et al. 1992), and in the sub-Himalaya (Suresh et al. 2007; Kumar et al. 2001). Having so far, in the former chapters, focused topically on alluvial fan processes, this final chapter presents the *regional approach*, i.e., a wider regional perspective of alluvial fans rendered along the Dead Sea-Arava Rift, Israel.

19.1 Setting

The *Dead Sea Transform* is a fault-bounded graben of the pull-apart, rhomb-shaped type, part of an intracontinental plate boundary between the Arabian plate and the Israel-Sinai subplate (Freund et al. 1970; Garfunkel 1997; Garfunkel et al. 2014). This north-south, about 1000 km long, transform is part of an oceanic opening and a crustal spreading that creates the Red Sea (Fig. 19.1). The rift is bounded by left-lateral strike-slip faults, along which the Arabian plate is pushed northward toward the Taurus-Zagros collision zone (Freund 1965; Wilson 1965). The transform is of Cenozoic origin, formed when the African-Arabian continent broke up. It has accumulated 105 km horizontal displacement (Freund et al. 1970) and remains the center of seismic activity to the present day (Freund et al. 1968; Zak and Freund 1966; Garfunkel et al. 1981; Bowman and Gerson 1986). Within the boundaries of Israel, the Dead Sea Transform is about 500 km long from south to north. This review focuses on the deepest continental segment of the transform, stretching from the Dead Sea through the Arava Rift Valley down to the Gulf of Elat.

Different issues have been the foci of alluvial fan studies along the Dead Sea-Arava transform. In the Dead Sea area, the drop of Lake Lisan and the rapid fall of the Dead Sea formed a complex fan morphology where the effect of a lowering base-level became an important issue. Southward along the Arava Rift, the alluvial fans served mainly as surfaces for recording morpho-tectonic fingerprints, monitoring normal and sinistral tectonic movements, and estimating their rates.

19.2 Lake Lisan and the Dead Sea

The Dead Sea area (Figs. 19.1 and 19.2), the lowest global topography, is hyperarid, with 50 mm mean annual precipitation and wide annual variations. The rift is flanked on the east by the Jordanian Edom Highlands which rise to 1700 m and indicate that the major uplift is on the east, while on the west, the altitudes rise only

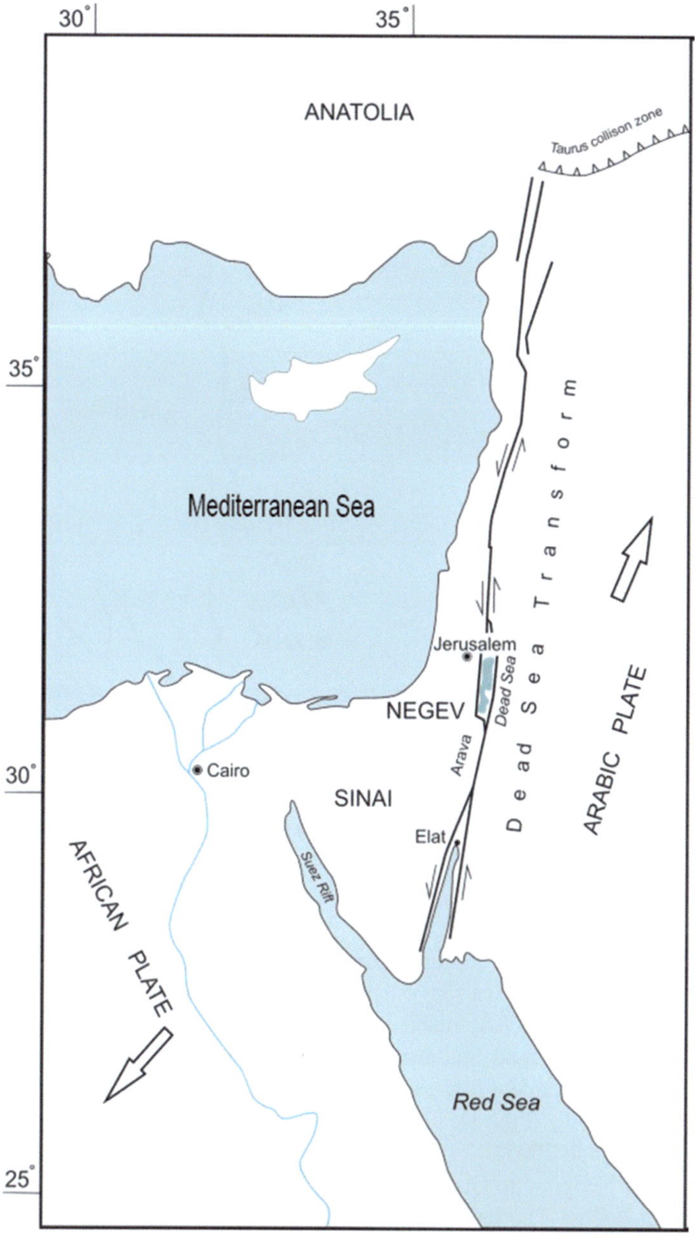

Fig. 19.1 Tectonic setting of the Dead Sea Transform between the Red Sea and the Taurus collision zone in Anatolia. The transform is marked by the Elat Gulf and the south-north strike-slip fault system

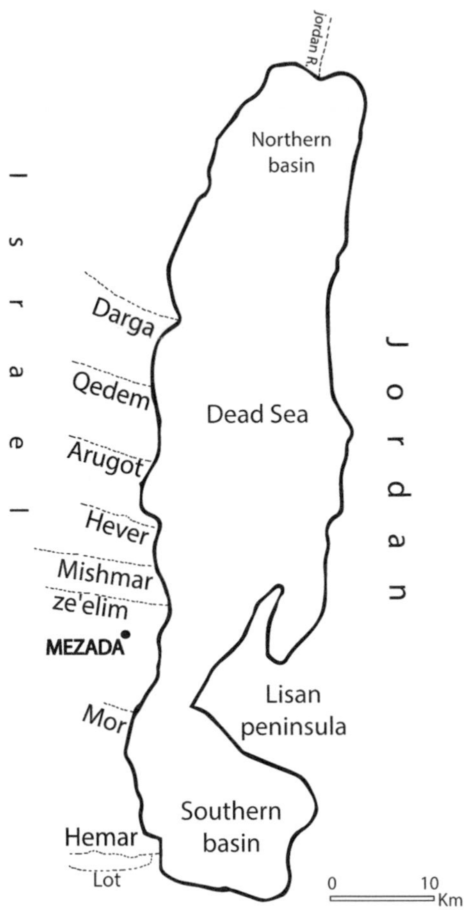

Fig. 19.2 Location map of the Dead Sea area

to 800–1000 m with mean annual precipitation exceeding 600 mm. Mean annual temperatures at the Dead Sea range from 15 °C in winter to 35 °C in summer, but surface soil temperatures may reach over 50 °C. Being essentially dry all but a few days of the year, the Dead Sea area demonstrates a typical hyperarid flash flood regime with very infrequent flow events. Steep channels slope 0°49′–1°25′ from the western margins toward the Dead Sea, and flows attain high velocities of 4–6 m/s (Cohen and Laronne 2005).

Throughout Neogene to Quaternary times, the Dead Sea basin was occupied by several water bodies. The last large lake, the precursor of the Dead Sea, was *Lake Lisan* of the last glacial period, from 70 to 14 ka BP (Begin et al. 1974, 2004; Haase-Schramm et al. 2004; Klinger et al. 2003; Neev and Emery 1967). At its peak, around 27,000 cal. years BP (Bartov et al. 2002), the lake stood at ~150 m below mean sea level and stretched along 240 km, from the Sea of Galilee to the northern Arava. Lake Lisan peak stage is marked by relict beach ridges in the northern Arava (Bowman and Gross 1992) as well as by the highest beach terraces on the Jordanian side (Ghazleh and Kempe 2009). The lake deposits – the *Lisan Formation* – are beautifully laminated with alternating layers of white aragonitic chalk, fragmental calcite, and dolomite.

The *Dead Sea*, the successor of Lake Lisan, is a Holocene hypersaline dense lake (1.23 g/cc). Its level, which is the lowest surface level on earth, dropped to −432 m below MSL in 2017. The Dead Sea is the terminal lake for the 43,000 km^2 drainage system of the Jordan River. The negative water balance of the Dead Sea during the twentieth century and its ever-growing shrinking are mainly due to the increased diversion of water from its drainage basin for watering in Jordan and in Israel and also due to damming of tributaries in Syria. In addition, 40% of the lake level drop is pumped from its deep north basin to evaporation pans in the south basin by the potash plants operating both in Israel and in Jordan. Thus, the recent extreme drop in the water level of the Dead Sea is largely a man-made event.

19.3 The Terminal Retreat of Lake Lisan

During its high stand, Lake Lisan inundated the pre-existing canyons that had been carved out of the Late Cretaceous to Tertiary limestones and dolomites in the flanking rift escarpments. The inundated canyons served as transport routes for fluviatile sediments, from the areal basins toward the rift. The final drying up of Lake Lisan, ~12–14 ka ago, left a series of 28 excellently preserved coastal terraces, evidence of standstills during the retreat of the lake, each showing a wave-cut tread and a riser (Bowman 1971, Fig. 19.3). In addition, Lake Lisan left a rich record of shoreline features that indicate the last overprint of the receding lake, including swash bars, beach ridges composed of well-sorted and rounded disc-shaped gravel, forming backsets and foresets, tar deposits, beach rock, and abandoned coastal cliffs, all excellently preserved along the western side of the Dead Sea as well as along the eastern Jordanian slopes (Ghazleh and Kempe 2009).

In order to follow the evolution of the alluvial fans in the Dead Sea area, the stages of Lake Lisan should be outlined, from the highest stand down to its final regressional level:

1. During the high lake level stand, fluviatile bedload was flushed through the inundated canyons to the rift, to become deposited as *sublacustrine fans* in front of the fault escarpment. The sublacustrine fans are the highest and oldest in the Dead Sea depression (Figs. 19.3 and 19.4). They are small, amounting to 2% of their source area, compared to 10% up to more than 100% of the source area in case of subaerial fans.
2. When the lake level dropped and the upper part of the sublacustrine fans became exposed with major parts still underwater, the fans became *Gilbert-type fans*. They typically display steeply inclined (24–34°) *foresets* consisting of planar couplets of clast-supported cobbles and small boulders alternating with granule-pebbly units. The foreset beds are capped with gentle dipping (<1°) fluviatile *topset* units. Fine exposures of Gilbert-type fans can be observed in front of the

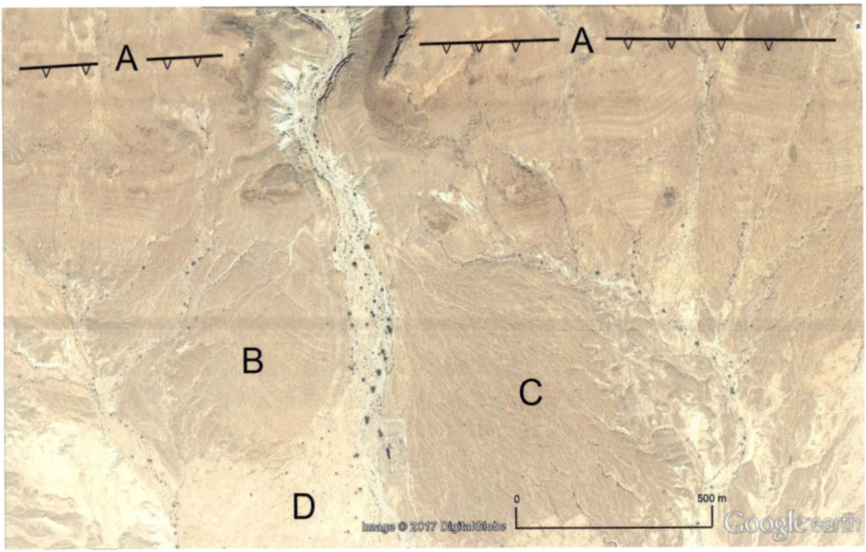

Fig. 19.3 Aerial view of the alluvial fan of the Mishmar creek west of the Dead Sea: (**a**) the rift escarpment with Lisan raised coastal terraces. (**b**) The residual smooth, unentrenched sublacustrine fan surface with contour-like beach ridges. Most of the sublacustrine fan had been eroded. (**c**) The post-Lisan alluvial fan. (**d**) The apex of the recent active fan. (Photo Google Earth 2017)

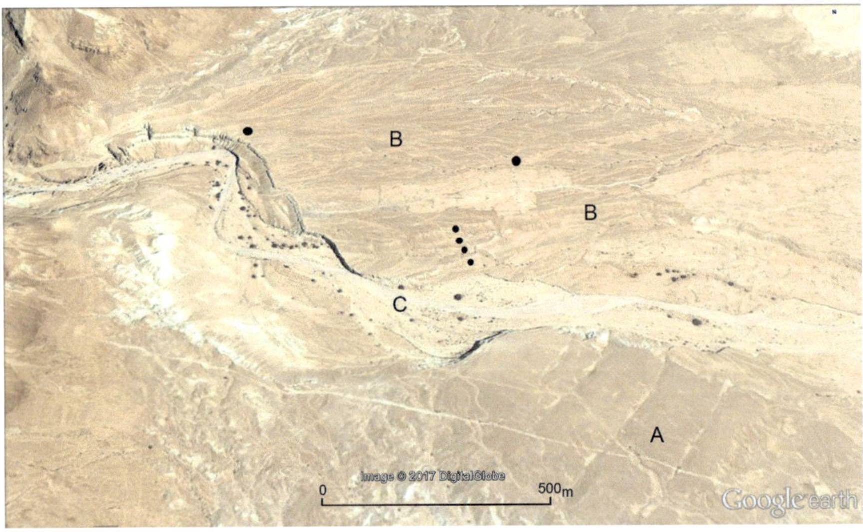

Fig. 19.4 Oblique view of the Ze'elim alluvial fan: (**a**) A residual of the initial sublacustrine smooth fan surface with its beach ridges. (**b**) The fan terraces. Some of the risers of the fan terraces are indicated. Braided channels on the treads of the fan terraces reveal the paleo-surface drainage. (**c**) The entrenched through-fan active channel. Compare to Fig. 19.5. (Photo Google Earth 2017)

Mor (Bowman 1990) and the Qedem canyons. Gilbert-type fans have also been deposited by other creeks during historical higher stands of the Dead Sea.

3. When the drop of the lake level continued, only the distal front of the fans kept contact with the receding lake, transforming the fans into *fan deltas*.
4. When Lake Lisan and the Dead Sea dropped rapidly, the progradation of part of the fans lagged behind, and the fans become completely disconnected from the regressing lakes. A recent example is the nearshore area of the Ze'elim (see front cover). The fans accumulated entirely in a subaerial environment as pure *alluvial fans*, often showing a *telescopic pattern*. No *delta* was ever deposited along the Lisan or Dead Sea margins, except the delta of the Jordan River as part of the axial drainage system.

The fans along the western Dead Sea shore are lined up in a narrow zone, up to 6 km wide, between the rift escarpment and the receding Dead Sea, within an altitudinal range of 150–432 m below MSL (Bowman 1974, 1997). Following the drop of Lake Lisan and the recent ongoing retreat of the Dead Sea, the Lisan Formation is exposed in interfan areas and lacustrine mudflats show in front of the alluvial fans (Neev and Emery 1967; Begin et al. 1974; Machlus et al. 2000; Bookman et al. 2006). In the first study devoted to fans in the Dead Sea area, Sneh (1979) recognized four types of facies: (1) conglomerate facies, (2) sandstone facies, (3) interlayered (rhythmic) sand/mud facies, and (4) laminated chalk facies. Fan sequences were also observed by Amit and Gerson (1986), Bowman (1988), and Frostick and Ried (1989).

19.4 Fan Incision

The receding and fluctuating Lisan and Dead Sea lake levels caused cyclic interfingering of lacustrine, shallow nearshore, beach, alluvial fan, and talus deposits. Following the drying up of Lake Lisan, the sublacustrine fans became gradually exposed and subject to incision. Vertical and lateral entrenchment started to cut into the exposed fans and in some cases stripped away significant parts, as those of the Mishmar and the Ze'elim creeks (Figs. 19.3 and 19.4). Deep entrenched valleys left the higher surfaces of the fans inactive. The lateral vector of the fan entrenchment commonly left behind fan terraces. The alluvial fans of the Dead Sea area are unique and differ from those of the Arava Valley by their high number of terraces, 18 terraces in the Hemar fan (drainage basin 326 km^2), 14 terraces in Ze'elim fan (drainage basin 236 km^2), and 12 in the Lot fan (drainage basin 20 km^2), compared to no more than 5 fan terraces reported in the fans of the southern Arava Rift Valley, reflecting the limited altitudinal range available there for incision. The mean slopes of the Dead Sea fans, along the fans' surface, the terraces, and along the active channels of the Ze'elim, the Lot, and the Hemar creeks, range $0° 33'$–$3° 32'$.

The Ze'elim fan, the largest along the western shore of the Dead Sea, provides a good example of the fan morphology and the entrenchment process in the Dead Sea area (Bowman 1978, 1988). Successive vertical incision and asymmetric southward lateral entrenchment carved in the Ze'elim fan 14 *fan terraces* (Figs. 19.4 and 19.5).

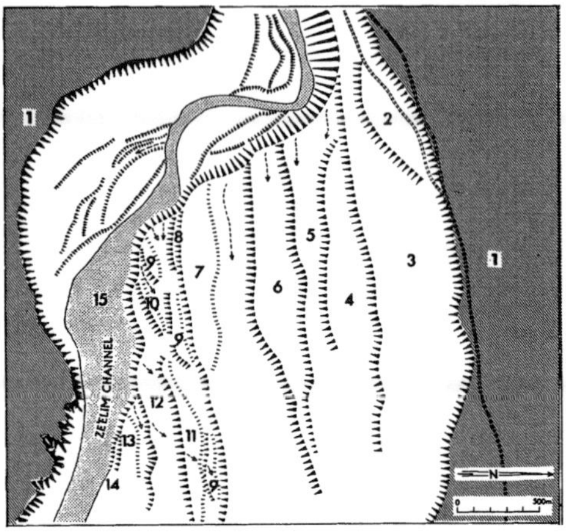

Fig. 19.5 Map of the 14 fan terraces of the Ze'elim. **1** indicates the residual sublacustrine fan surface. The risers are dying out basinward

This lateral entrenchment on the Ze'elim fan destroyed most of its initial sublacustrine fan surface, except at the most southern (Fig. 19.4a) and northern margins, where the initial smooth sublacustric fan surface with its beach ridges have been preserved. The fan terraces are capped by a gravel veneer of limestone, dolomite, and chert up to 5 m thick. Their surface shows a rugged bar-and-swale morphology with bar relief up to 2 m-high and 20 m-wide swales. The Ze'elim fan terraces have been chosen to examine soil changes within the Holocene and to find out what are the rates of change of the different soil properties (Amit and Gerson 1986).

19.5 The Lowering Base-Level Effect

The level of Lake Lisan started to drop from its maximal elevation at 150 m below mean sea level at ~14,000 YBP. Recently, the measured elevation of the Dead Sea dropped from −388 m in 1896 to −432 m in 2017 and still continues to fall rapidly, reaching records of over 1 m/year. Such a dramatic base-level fall made the Dead Sea area with its soft unconsolidated sediments a unique field laboratory for studying the control of a lowering base-level on alluvial fans.

It was tempting to explore and model the incision of the Dead Sea fans as a response to the continuous and dramatic drop of Lake Lisan and the Dead Sea levels. The proximity of the many well-preserved fan terraces to the extensive series of relict beach terraces strengthened the hypothesis that here we are given an example of fan terrace formation controlled by a receding base-level. The view of base-level fall as the trigger for entrenchment has been hypothetically inferred by Quennell (1958). Similarly, knickpoints in longitudinal stream profiles have been associated with the lowering of the base-level by Willis (1928), Nir and Ben-Arie (1965), and by Sneh (1979). The late Prof. Luna Leopold, striving to share his experience and advice the author, years ago when a PhD candidate, thought that the Dead Sea area was a prime case for studying the effect of base-level-lowering and pointed to the relative meager knowledge yet gained on this topic in the USA.

There are major reasons to deny a base-level control in the fan environment along the Dead Sea:

1. The surfaces of all the fans, including of the fan terraces, except the sublacustrine highest which are irrelevant, do not bear any evidence of wave-cutting or beach depositions, i.e., they show no evidence of a nearby Paleo-shoreline. Moreover, all the fans' front areas, where the fan surfaces or terraces die out and a controlling base-level could be expected, do not bear any evidence – morphological or sedimentary – of a raised shoreline which could have acted as a controlling base-level.
2. The different numbers of fan terraces on the various fans rule out a single simultaneous regional controlling base-level. If controlled by a regional base-level, the number of fan terraces should be similar.
3. The different numbers of Lisan raised beaches (about 40; Bowman 1971) and of the fan terraces (12–18) suggest that we are dealing with two independent systems.
4. The longitudinal profiles of the fan terraces of the Lot and the Ze'elim creeks converge downstream, indicating merging of the surfaces (Bowman 1988). Converging terraces usually characterize flights of fan terraces (Hooke 1972) but do not indicate a lowering base-level.
5. Longitudinal profiles of recent channels and of terrace remnants that have been undoubtedly controlled by the recent rapid fall of the Dead Sea level do not show convergence but tend toward parallelism or divergence (Moshe et al. 2008; Storz-Peretz et.al. 2011), suggesting that longitudinal converging fan terraces do not indicate a lowering base-level control.

The Lake Lisan level fall must have been fast, causing fan progradation to lag and loose contact with the dropping base-level. The rapid level drop *allowed entrenchment but did not cause it*. The fall of the Lake Lisan level provided space for fan entrenchment and for progradation that were activated by floods *from the apex down-fan*. The receding lake levels played only a passive role, i.e., permitting entrenchment without transmitting headward any significant control.

The recent active response to the extreme base-level drop is best explored at the front of fan deltas which were recently exposed along the receding Dead Sea. The rates of incision in the channels, within a few hundred meters from the nowadays receding shoreline, were measured lately as 0.4–0.8 m/year. These rates are extreme in global terms, i.e., compared to 0.02–0.05 m $year^{-1}$ maximum rates of incision in alluvium and in silt beds reported by Phillips et al. (2005) and by Simon and Rinaldi (2006). The extreme channel entrenchment by the Dead Sea shoreline excavates nowadays steep channels with longitudinal slope gradients of 1.7–3.4°, 0.6–3.2°, and 5.5° at the Arugot, Ze'elim, and Qedem creeks, respectively. The active channel of the Qedem creek at the nearshore area has been temporarily observed in a "hanging" position, i.e., the channel terminated on top of a small coastal scarp, disconnected from and uncontrolled by the retreating shoreline. Subsequent floods eradicated the coastal scarp and readjusted the channel to the receding Dead Sea level (Storz-Peretz et al. 2011). Flood frequency in the area averages 0.78 per year with 0.7 $year^{-1}$ standard deviation (Ben-Zvi 2006), meaning that the headward transmission of the base-level effect had not been blocked over periods longer than a few single dry years.

The rapid vertical incision is accompanied by some *channel widening*, forming a trapezoidal cross section with steep channel banks, typified by tension cracks and by failed slabs. Such widening caused failure of a bridge over the Arugot Creek, located within a few hundred meters from the shoreline. However, this channel widening does not mark the final evolutional stage of cross sections, conceptualized by models, when widening counteracts the deepening trend and becomes a major mechanism dissipating the excess stream power. Being controlled by the continuous rapid base-level fall, the channels in the Dead Sea area do not widen considerably but are entrapped in the vertical incisional mode which prevents the evolutionary progress toward a significant wider, graded cross-sectional shape. Significant channel widening is retarded by the imposed continuous, rapid base-level drop and does not become an efficient negative feedback to deepening (Bowman et al. 2010). The Dead Sea case is of value for predicting evolutional scenarios of cross sections under similar conditions of a long-term, rapid base-level fall.

The effect of a base-level fall that exposes a steep slope had been studied by experiments that simulated the evolution of drainage networks upslope of a falling base-level (Pelletier 2003) often focusing on the shoreline drop below the shelf break (Koss et al. 1994). The experiments demonstrated that at this stage channels began to form by migration of knickpoints headward, one past the other, while performing captures and channel piracy. This incision and growth of channels headward resulted in the experimental studies in formation of a dendritic pattern. The rapid fall of the Dead Sea level, attaining 24 m during the 1945–2002 period, exposed large coastal areas in front of alluvial fans. Visual interpretation of orthophotos showed that the drainage pattern changed when flow prograded from the fans to the exposed areas in their front. To examine the *connectivity* and the *coupling* (Brunsden and Thornes 1979; Harvey 2002) seems here most relevant. The drainage organization in the Dead Sea area displays transformation from a diverging, bifurcating network on the active fans into converging, dendritic systems on the newly emerged areas in front of the alluvial fans (Bowman et al. 2011; Fig. 19.6 and front cover).

19.6 Alluvial Fans of the Arava Rift Valley

The Arava Rift Valley is the continental continuation of the Gulf of Elat, which is the northward extension of the Red Sea along the Dead Sea Transform (Figs. 19.1 and 19.7). The northern Arava drains into the Dead Sea, whereas the southern Arava drains mainly into closed basins and into the Gulf of Elat with the water divide at the mid-Arava zone. Both margins of the Arava Rift Valley are bordered by alluvial

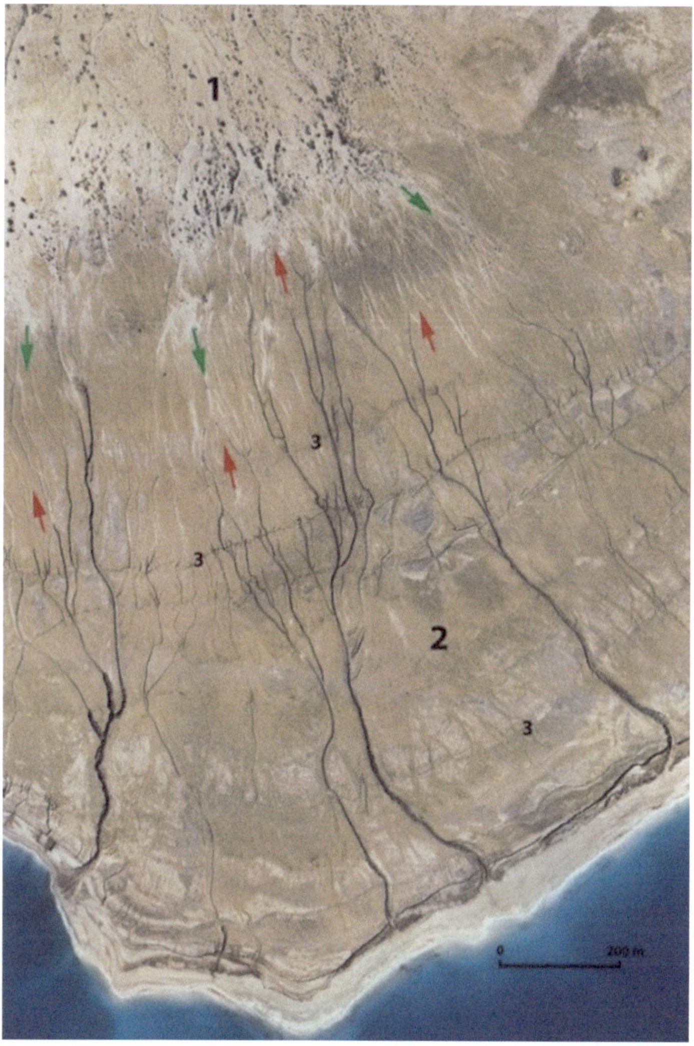

Fig. 19.6 Drainage evolution in front of the Ze'elim alluvial fan (1) following base-level fall: the active propagating and diverging channels at the distal front of the Ze'elim fan (green arrows) transform on the recently exposed smooth lake bed (2) with its raised shorelines (3) into a converging dendritic system (red arrows), entrenching headward and capturing more runoff, which is a sort of positive feedback, i.e., enhancing the change from aggradation on the fan to entrenchment on the emerged areas

fans with the main transform fault running along the sharp front of the eastern rift margins (Fig. 19.8). The uplifted eastern rift shoulder, the Edom Mountains, is higher than the western shoulder, and the alluvial fans along the eastern side are often much larger, reflecting the structural asymmetry of the Arava Valley. The continental segment of the Elat fault emerges from the gulf and heads northward along

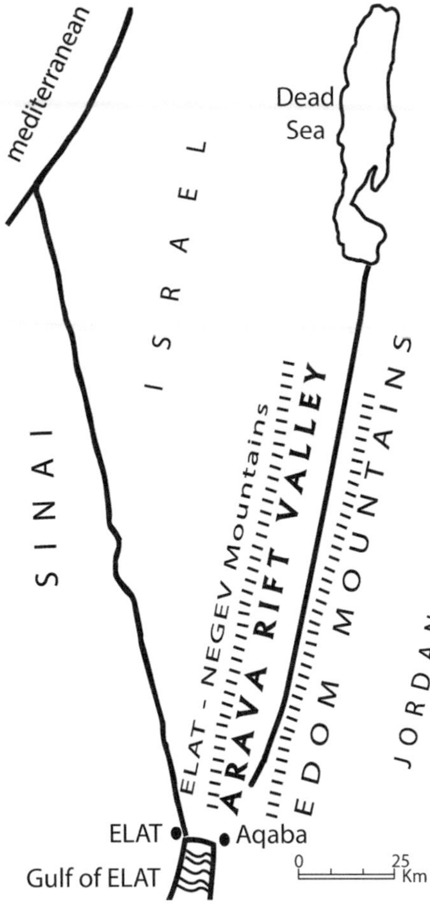

Fig. 19.7 Location map of the Arava Rift Valley, between the Dead Sea and the Gulf of Elat. Note the eastern rift margins of the Edom Mountains and the western Elat-Negev rift shoulders

the western rift margin, into the southern Arava. The climate in the region is extremely arid, with summer temperatures rising over 40° and a mean annual rainfall of 30 mm.

The alluvial fans along the southern and northern Arava valley are of Late Pleistocene and Holocene ages. In the central Arava, most fans range from Early to Late Pleistocene age indicating a limited subsidence. The age of the uppermost fan surface of the Shehoret fan (Fig. 19.8) composed of limestone, chert, sandstone, granite, schist, and diorite, based on IRSL, has been found to be in the 60–75 ka range (Porat et al. 1997). Other ages of fan surfaces in the Arava ranged between 64 and 129 ka, partly expressing inheritance (Le Béon et al. 2012).

The uppermost abandoned Pleistocene fan surfaces in the southern Arava typically show a reg soil with a gravel-free gypsic B horizon, shattered gravel, and a petorosalic C horizon (Amit et al. 1995). No coalescing of fans is observed along the Arava margins, nor along the Holocene fans of the Dead Sea, and interfan areas are common. The *bajada coalescing stage* has not been achieved yet.

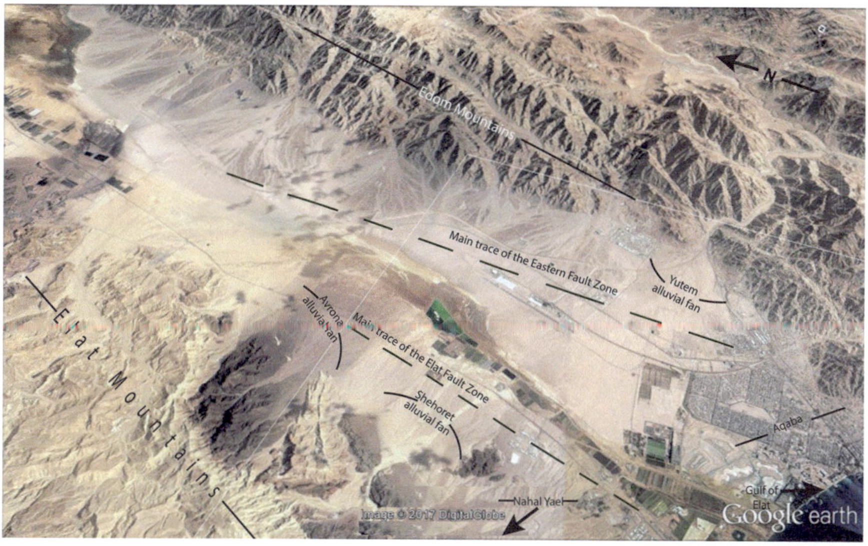

Fig. 19.8 Oblique view from the Gulf of Elat northward toward the Arava Rift Valley. Note the alluvial fans along the rift margins and the rift-parallel fault zones. These alluvial fans display in the field Late Pleistocene and Holocene displaced surfaces, grabens, push-ups, and scarp lineaments (shown in Fig. 19.9). Following Zilberman et al. (2005), Garfunkel et al. (1981), and Amit et al. (2002)

Regarding nourishment of alluvial fans, of value is the study carried out in the small (0.6 km^2), intensively instrumented hyperarid basin of Yael creek north to Elat (Fig. 19.8) by Schick and Lekach (1993) and Clapp et al. (2000). They reported that short-term sediment yield exceeds long-term sediment production by 53–86%. Such significant difference between the rates of sediment generation and sediment yield indicates trapping of colluvium. Basins are currently evacuating toward their fan sediments stored during periods of more rapid sediment generation.

In the last 1000 years, the Dead Sea Transform experienced destructive earthquakes of magnitude 5.8–6.2. A most active segment is located in the Gulf of Elat where the last severe earthquake in 1995 reached M 7.2 with the epicenter 70 km south of the cities of Elat and Aqaba. Following the seismic activity along the Arava Rift Valley, the alluvial fans became the focus of research as surfaces that clearly reveal tectonic displacements. Alluvial fans displaced by several en echelon faults and by small rhomb-shaped grabens, whose geometry indicates a left-lateral slip, have already been reported by Zak and Freund (1966). Rupture of fans and deformation of Late Pleistocene and Holocene alluvial fans, including playa deposits, have been reported by Bowman and Gerson (1986), Garfunkel et al. (1981), and Gerson et al. (1993). The displacements have often a surface expression of scarps or shallow saddles across the fan surface, arranged in straight lineaments striking north (Fig. 19.9). Trenching enriched the morpho-tectonic evolutional picture by revealing multiple faulting events (Amit et al. 1995, 2002). Proximity in time between

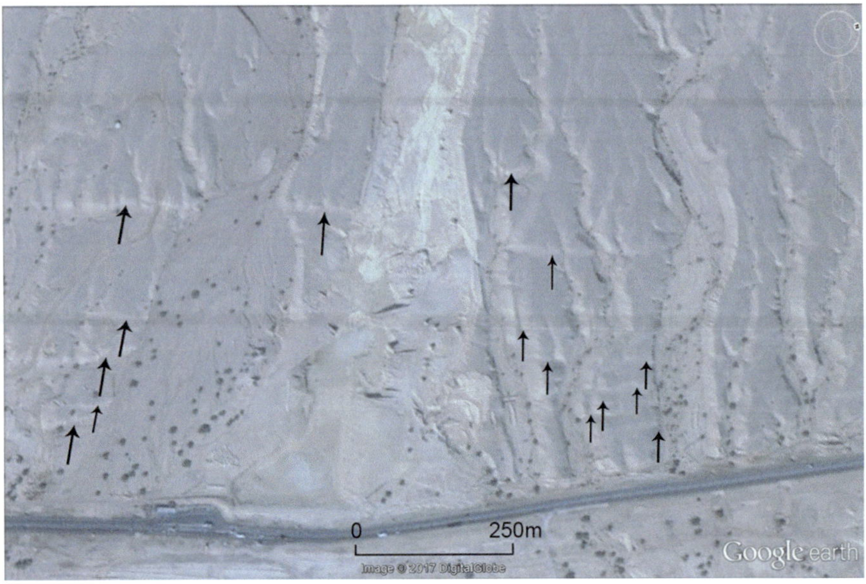

Fig. 19.9 A flight of parallel fault scarps displacing the Shehoret fan surface in the southern Arava fault zone, north of Elat (Fig. 19.8). Post-tectonic undisplaced surfaces are clearly observable. The scarps have been intensively trenched (Amit et al. 1995, 1996; Zilberman et al. 2005). The uppermost fan surface was dated 56 ka. (Photo Google Earth 2017)

tectonic events is evident in the trenches from the absence of pedogenic features at contacts between the deposited units.

Alluvial fans that are detached, sometimes by a depression, from their sourcing drainage basin, indicate along the Arava a left-lateral slip and assist in determining its amount (Garfunkel et al. 1981; Ginat et al. 1998). Along the eastern margins of the Arava Rift, sharp linear edges of alluvial fans and offsets of fans relative to their feeding catchments have been reported by Le Béon et al. (2012). Restorations tend to favor slip rates of 5–7 mm/year for the last ~100 ka that are consistent with the present-day GPS slip rate of 3.5–6.3 mm/year (Garfunkel and Ben Avraham 2001).

References

Amit R, Gerson R (1986) The evolution of Holocene Reg (gravelly) soils in deserts – an example from the Dead Sea region. Catena 13:59–79

Amit R, Harrison JBJ, Enzel Y (1995) Use of soils and colluvial deposits in analysing tectonic events – the southern Arava Rift, Israel. Geomorphology 12:91–107

Amit R, Harrison JBJ, Enzel Y, Porat N (1996) Soils as a tool for estimating ages of Quaternary fault scarps in a hyperarid environment—The southern Arava valley, The Dead Sea Rift, Israel. Catena 28(1–2):21–45

Amit R, Zilberman E, Enzel Y, Porat N (2002) Paleoseismic evidence for time dependency of seismic response on a fault system in the southern Arava Valley, Dead Sea rift, Israel. GSA Bull 114(2):192–206

Bartov Y, Stein M, Enzel Y, Agnon A, Reches ZE (2002) Lake levels and sequence stratigraphy of Lake Lisan, the late Pleistocene precursor of the Dead Sea. Quat Res 57(1):9–2
Begin ZB, Ehrlich A, Nathan Y (1974) Lake Lisan: the pleistocene precursor of the Dead Sea. Geological Survey of Israel, Jerusalem
Begin ZB, Stein M, Katz A, Machlus M, Rosenfeld A, Buchbinder B, Bartov Y (2004) Spatial trends in the late quaternary Lisan formation (Dead Sea rift) and their climatic implications. Quat Sci Rev 23:1627–1636
Ben-Zvi A (2006) A regional model for floods in the Western Dead Sea channels and its application to Nahal Qedem. Israel Hydrological Survey Report 2006/3, 10 pp (in Hebrew)
Bookman R, Bartov Y, Enzel Y, Stein M (2006) Quaternary lake levels in the Dead Sea basin: two centuries of research. Geol Soc Am Spec Pap 401:155–170
Bowman D (1971) Geomorphology of the shore terraces of the late Pleistocene Lisan Lake (Israel). Palaeogeogr Palaeoclimatol Palaeoecol 9(3):183–209
Bowman D (1974) Geomorphology of river terraces on the western bank of the Dead Sea, International Symposium: Geomorphic Processes in Arid Environments, Field Program. The Hebrew University of Jerusalem, Jerusalem, pp 40–51
Bowman D (1978) Determination of intersection points within a telescopic alluvial fan complex. Earth Surf Process 3:265–276
Bowman D (1988) The declining but non-rejuvenating base level-The Lisan Lake, the Dead Sea area, Israel. Earth Surf Process Landf 13(3):239–249
Bowman D (1990) Climatically triggered Gilbert-type lacustrine fan deltas, the Dead Sea area, Israel. Spec Publ Int Assoc Sedimentol 10:273–280
Bowman D (1997) Geomorphology of the Dead Sea western margins. In: Niemi TM, Ben-Avraham Z, Gat JR (eds) The Dead Sea-the Lake and its setting, Oxford Monographs on Geology and Geophysics No. 36. Oxford University Press, New York, pp 217–225
Bowman D, Gerson R (1986) Morphology of the latest Quaternary surface-faulting in the Gulf of Elat region, eastern Sinai. Tectonophysics 128(1–2):97–119
Bowman D, Gross T (1992) The highest stand of Lake Lisan: ~150 meter below MSL. Israel J Earth-Sci 41(2–4):233–237
Bowman D, Svoray T, Devora S, Shapira I, Laronne JB (2010) Extreme rates of channel incision and shape evolution in response to a continuous, rapid base-level fall, the Dead Sea, Israel. Geomorphology 11(3):227–237
Bowman D, Devora S, Svoray T (2011) Drainage organization on the newly emerged Dead Sea bed, Israel. Quat Int 233(1):53–60
Brunsden D, Thornes JB (1979) Landscape sensitivity and change. Trans Inst Br Geogr New Ser 4:463–484
Clapp EM, Bierman PR, Schick AP, Lekach J, Enzel Y, Caffee M (2000) Sediment yield exceeds sediment production in arid region drainage basins. Geology 28(11):995–998
Cohen H, Laronne JB (2005) High rates of sediment transport by flashfloods in the Southern Judean Desert, Israel. Hydrol Process 19(8):1687–1702
Freund R (1965) A model of the structural development of Israel and adjacent areas since Upper Cretaceous times. Geol Mag 102(3):189–205
Freund R, Zak I, Garfunkel Z (1968) Age and rate of the sinistral movement along the Dead Sea rift. Nature 220(5164):253–255
Freund R, Garfunkel Z, Zak I, Goldberg M, Weissbrod T, Derin B, Bender F, Wellings FE, Girdler RW (1970) The shear along the Dead Sea rift. Philos Trans R Soc Lond A Math Phys Sci 267(1181):107–130
Frostick LE, Reid I (1989) Climatic versus tectonic controls of fan sequences: lessons from the Dead Sea, Israel. J Geol Soc 146(3):527–538
Garfunkel Z (1997) The history and formation of the Dead Sea basin. In: Niemi TM, Ben Avraham Z, Gat JR (eds) The Dead Sea: the Lake and its setting. Oxford University Press, New York, pp 36–56

Garfunkel Z, Ben-Avraham Z (2001) Basins along the Dead Sea transform. In: Ziegler et al (eds) Peri-Tethys Memoir, Peri-Tethyan Rift/Wrench Basins and Passive Margins, vol 6. Mem. Mus. Natl. Hist. Nat., Paris, pp 607–627
Garfunkel Z, Zak I, Freund R (1981) Active faulting in the Dead Sea. Tectonophysics 80(1–4):1–26
Garfunkel Z, Ben-Avraham Z, Kagan E (eds) (2014) Dead Sea transform fault system: reviews. Springer, Netherlands
Gerson R, Grossman S, Amit R, Greenbaum N (1993) Indicators of faulting events and periods of quiescence in desert alluvial fans. Earth Surf Process Landf 18(3):181–202
Ghazleh SA, Kempe S (2009) Geomorphology of Lake Lisan terraces along the eastern coast of the Dead Sea, Jordan. Geomorphology 108(3):246–263
Ginat H, Enzel Y, Avni Y (1998) Translocated Plio-Pleistocene drainage systems along the Arava fault of the Dead Sea transform. Tectonophysics 284(1–2):151–160
Haase-Schramm A, Goldstein SL, Stein M (2004) U-Th dating of Lake Lisan (Late Pleistocene Dead Sea) aragonite and implications for glacial East Mediterranean climate change. Geochim Cosmochim Acta 68(5):985–1005
Harvey AM (1990) Factors influencing quaternary alluvial fan development in southeast Spain. Alluvial fans: a field approach. Wiley, Chichester, pp 247–269
Harvey AM (2002) Effective time scales of coupling within fluvial systems. Geomorphology 44:175–201
Harvey AM, Wigand PE, Wells SG (1999) Response of alluvial fan systems to the late Pleistocene to Holocene climatic transition: contrasts between the margins of pluvial Lakes Lahontan and Mojave, Nevada and California, USA. Catena 36(4):255–281
Hooke RL (1972) Geomorphic evidence for late-Wisconsin and Holocene tectonic deformation, Death Valley, California. Geol Soc Am Bull 83(7):2073–2098
Klinger Y, Avouac JP, Bourles D, Tisnerat N (2003) Alluvial deposition and lake-level fluctuations forced by Late Quaternary climate change: the Dead Sea case example. Sediment Geol 162(1):119–139
Koss JE, Ethridge FG, Schumm SA (1994) An experimental study of the effects of base-level change on fluvial, coastal plain and shelf systems. J Sediment Res 64(2):90–98
Kumar SA, Parkash B, Mohindra R, Thomas JV, Singhvi AK (2001) Quaternary alluvial fan sedimentation in the Dehradun valley piggyback basin, NW Himalaya: tectonic and palaeoclimatic implications. Basin Res 13(4):449–471
Le Beon M, Klinger Y, Mériaux AS, Al-Qaryouti M, Finkel RC, Mayyas O, Tapponnier P (2012) Quaternary morphotectonic mapping of the Wadi Araba and implications for the tectonic activity of the southern Dead Sea fault. Tectonics, 31(5), TC5003, https://doi.org/10.1029/2012TC003112
Machlus M, Enzel Y, Goldstein SL, Marco S, Stein M (2000) Reconstructing low levels of Lake Lisan by correlating fan-delta and lacustrine deposits. Quat Int 73:137–144
Moshe LB, Haviv I, Enzel Y, Zilberman E, Matmon A (2008) Incision of alluvial channels in response to a continuous base level fall: field characterization, modeling, and validation along the Dead Sea. Geomorphology 93(3):524–536
Neev D, Emery KO (1967) The Dead Sea: depositional processes and environments of evaporites. Geological Survey of Israel, Jerusalem
Nir D, Ben-Arie D (1965) Relicts of an intermediate terrace between the Ghor and the Zor in the Central Jordan Valley. Lake Tiberias–Kefar Rupin, Israel. Israel J Earth Sci 14:1–8
Pelletier JD (2003) Drainage basin evolution in the rainfall erosion facility: dependence on initial conditions. Geomorphology 53(1):183–196
Phillips JD, Slattery MC, Musselman ZA (2005) Channel adjustment of the lower Trinity River, Texas, downstream of Livingstone Dam. Earth Surf Process Landf 30:1419–1439
Porat N, Amit R, Zilberman E, Enzel Y (1997) Luminescence dating of fault-related alluvial fan sediments in the southern Arava Valley, Israel. Quat Sci Rev 16(3–5):397–402
Quennell AM (1958) The structural and geomorphic evolution of the Dead Sea rift. Q J Geol Soc 114(1–4):1–24

Schick AP, Lekach J (1993) An evaluation of two ten-year sediment budgets, Nahal Yael, Israel. Phys Geogr 14(3):225–238

Silva PG, Harvey AM, Zazo C, Goy JL (1992) Geomorphology, depositional style and morphometric relationships of Quaternary alluvial fans in the Guadalentin depression (Murcia, southeast Spain). Z Geomorphol 36(3):325–341

Simon A, Rinaldi M (2006) Disturbance, stream incision, and channel evolution: the roles of excess transport capacity and boundary materials in controlling channel response. Geomorphology 79:361–383

Sneh A (1979) Late Pleistocene fan-deltas along the Dead Sea rift. J Sediment Res 49(2):541–552

Sohn MF, Mahan SA, Knott JR, Bowman DD (2007) Luminescence ages for alluvial-fan deposits in Southern Death Valley: implications for climate-driven sedimentation along a tectonically active mountain front. Quat Int 166(1):49–60

Storz-Peretz Y, Bowman D, Laronne JB, Svoray T (2011) Rapid incision of a small, coarse and steep fan-delta in response to base-level fall: the case of Nahal Qedem, the Dead Sea, Israel. Earth Surf Process Landf 36(4):467–480

Suresh N, Bagati TN, Thakur VC, Kumar R, Sangode SJ (2002) Optically stimulated luminescence dating of alluvial fan deposits of Pinjaur Dun, NW Sub Himalaya. Curr Sci 82(10):1267–1274

Suresh N, Bagati TN, Kumar R, Thakur VC (2007) Evolution of Quaternary alluvial fans and terraces in the intramontane Pinjaur Dun, Sub-Himalaya, NW India: interaction between tectonics and climate change. Sedimentology 54:809–833

Willis B (1928) Dead Sea problem: rift valley or ramp valley? Geol Soc Am Bull 39(2):490–542

Wilson JT (1965) A new class of faults and their bearing on continental drift. Nature 207(4995):343–347

Zak I, Freund R (1966) Recent strike slip movement along the Dead Sea rift. Israel J Earth Sci 15:33–37

Zilberman E, Amit R, Porat N, Enzel Y, Avner U (2005) Surface ruptures induced by the devastating 1068 AD earthquake in the southern Arava Valley, Dead Sea rift, Israel. Tectonophysics 408(1):79–99

MIX
Papier aus verantwortungsvollen Quellen
Paper from responsible sources
FSC® C105338

If you have any concerns about our products,
you can contact us on
ProductSafety@springernature.com

In case Publisher is established outside the EU,
the EU authorized representative is:
Springer Nature Customer Service Center GmbH
Europaplatz 3, 69115 Heidelberg, Germany

Printed by Libri Plureos GmbH
in Hamburg, Germany